Lecture Notes in Computational Science and Engineering

93

Editors:

Timothy J. Barth
Michael Griebel
David E. Keyes
Risto M. Nieminen
Dirk Roose
Tamar Schlick

For further volumes:
http://www.springer.com/series/3527

Michael Bader • Hans-Joachim Bungartz
Tobias Weinzierl
Editors

Advanced Computing

Springer

Editors
Michael Bader
Hans-Joachim Bungartz
Tobias Weinzierl
Department of Informatics
Technische Universität München
Munich, Germany

ISSN 1439-7358
ISBN 978-3-642-38761-6 ISBN 978-3-642-38762-3 (eBook)
DOI 10.1007/978-3-642-38762-3
Springer Heidelberg New York Dordrecht London

Library of Congress Control Number: 2013950610

Math. Subj. Class. (2010): 35-06, 68U20, 68W10, 65Y05, 65K05, 65Y20, 76D05, 76S05, 68N30, 68U05

Printed on acid-free paper

Springer is part of Springer Science+Business Media (www.springer.com)

The most scientifically important and economically promising research frontiers in the 21st century will be conquered by those most skilled with advanced computing technologies and computational science applications.

From: "Computational Science: Ensuring America's Competitiveness", President's Information Technology Advisory Committee (PITAC), 2005

Preface

Advanced Computing

The opening statement of this book, one of the key conclusions of the 2005 PITAC report, labels *Advanced Computing* and *computational science applications* as fields of research with high scientific and economic relevance. Today, computational science and engineering (CSE) has become a highly interdisciplinary, challenging, and thriving scientific domain of increasing importance and visibility in research, development, and education. Its essence involves modelling and computations – in particular (numerical) simulations – on computer systems ranging from standard PC to supercomputers. Thus, CSE applications have always been a driving force of computing technologies, especially progressing the field of supercomputing.

However, regarding *advanced computing technologies*, an increasing gap is observed between what should be possible in theory – due to recent advancements in algorithms, hardware, and networks – and what can really be achieved in practice. While increase of hardware performance (Moore's law) and considerations on complexity and accuracy of algorithms were driving this field over the last decades, new limiting factors are now encountered: hardware awareness and ubiquitous parallelism due to many-core systems (to name only two such issues) transform computing into a multifaceted process spawning modelling, algorithmics, parallel programming, and even hardware design; simulation data are produced to an extent and resolution which makes their mere handling, their analysis and, thus, the extraction and representation of relevant information a serious challenge – especially if to be done interactively; the development of software for high-performance computing (HPC), ranging from specialised single-application codes to general problem-solving environments, has thus become an increasingly complex process, comparable to what we have known for decades from other fields, and it needs sophisticated language and tool support as well as a substantial professionalisation.

Hence, to be able to exploit future (super)computers at their full potential and to expand the research frontiers mentioned above, a concerted effort is necessary in *Advanced Computing* – which is thus defined as HPC together with all its enabling technologies crucial to tackle the old and new bottlenecks. Such an endeavour must combine computing technologies and applications and bring together the HPC community as well as application specialists with experts from the enabling technologies.

The Munich Centre of Advanced Computing

Starting from these considerations, in 2008, the *Munich Centre of Advanced Computing* (MAC) has been established as an independent research consortium and common roof for computing-related activities of research groups at Technische Universität München (TUM) – together with several partner institutions, such as the Leibniz Supercomputing Centre, the Max-Planck Institute for Astrophysics, or the Department of Geophysics of Ludwig-Maximilians-Universität München. On the funding side, MAC started on two pillars: eight projects cofinanced by the State of Bavaria and by TUM and two projects established in the framework of a strategic partnership between TUM and KAUST, the King Abdullah University of Science and Technology, in Jeddah, Saudi Arabia. Fostering a structured programme for PhD studies, MAC is participating in TUM's International Graduate School of Science and Engineering (IGSSE). All MAC PhD students thus engaged in IGSSE's research training programme – one of the cornerstones being a 3-month research visit at an international partner institute.

Articles in These Proceedings

The present proceedings volume combines 11 articles that review research work and results from the MAC project teams. The first two articles, by Schillinger et al. and Benk et al. focus on the Finite Cell Method and on Immersed Boundary Methods, respectively, and thus on the question how to efficiently and accurately discretise problems with complicated domain boundaries on structured (adaptive) Cartesian grids, in order to simplify mesh generation and improve computational performance. Cai et al. discuss how micro- and macroscale modelling can be combined for accurate simulation of porous media flow, with CO_2-sequestration as the intended application. Simon and Ulbrich present their work on optimal control problems for this application area.

As an example for the development of novel mathematical algorithms, Böhm and Ulbrich present a Newton-CG method for full-waveform seismic inversion. Roderus et al. demonstrate how advancing both mathematical and computational algorithms can lead to substantial improvement of the performance of established computational science codes on HPC platforms – their focus being on density functional theory as application. Kraja et al. in their article, put the HPC platforms first in the development chain and discuss how to design HPC architectures for given applications; here, on-board processing for SAR image reconstruction on spacecrafts is considered. As simulation tasks and codes, respectively, become more and more complex, systematic software engineering has become imperative – Li et al. address this question via a requirements engineering approach for a software package in computational seismology.

The following two articles both address interactive simulations: while by Knežević et al. present an interactive computing framework for engineering applications, in particular, Benzina et al. introduce a framework that uses surrogate models based on sparse grids to allow interactive handling of high-dimensional simulation data. Both projects used the FRAVE – a flexible, reconfigurable visualisation environment installed at MAC – for their studies. Tönnis et al. as our last article in these proceedings, focus on the developers and users of such interactive simulation and visualisation environments: they present experiences and findings regarding ergonomic aspects of setting up and working with the FRAVE.

Taken altogether, the articles thus provide an overview of research in MAC from 2008 to 2012 and cover (nearly) the entire simulation pipeline.

Munich, Germany
May 2013

Michael Bader
Hans-Joachim Bungartz
Tobias Weinzierl

Contents

A Review of the Finite Cell Method for Nonlinear Structural Analysis of Complex CAD and Image-Based Geometric Models

Dominik Schillinger, Quanji Cai, Ralf-Peter Mundani, and Ernst Rank

Abstract The finite cell method (FCM) belongs to the class of immersed boundary methods, and combines the fictitious domain approach with high-order approximation, adaptive integration and weak imposition of unfitted Dirichlet boundary conditions. For the analysis of complex geometries, it circumvents expensive and potentially error-prone meshing procedures, while maintaining high rates of convergence. The present contribution provides an overview of recent accomplishments in the FCM with applications in structural mechanics. First, we review the basic components of the technology using the p- and B-spline versions of the FCM. Second, we illustrate the typical solution behavior for linear elasticity in 1D. Third, we show that it is straightforward to extend the FCM to nonlinear elasticity. We also outline that the FCM can be extended to applications beyond structural mechanics, such as transport processes in porous media. Finally, we demonstrate the benefits of the FCM with two application examples, i.e. the vibration analysis of a ship propeller described by T-spline CAD surfaces and the nonlinear compression test of a CT-based metal foam.

Keywords Embedded/fictitious domain methods • finite cell method • large deformation solid mechanics

D. Schillinger (✉)
Lehrstuhl für Computation in Engineering, Department of Civil Engineering and Surveying, Technische Universität München, Arcisstr, 21, 80333 München, Germany

Institute for Computational Engineering and Sciences, The University of Texas at Austin, 201 East 24th Street, Austin, TX 78712, USA
e-mail: dominik@ices.utexas.edu

Q. Cai · R.-P. Mundani · E. Rank
Lehrstuhl für Computation in Engineering, Department of Civil Engineering and Surveying, Technische Universität München, Arcisstr, 21, 80333 München, Germany
e-mail: cai@bv.tum.de; mundani@bv.tum.de; rank@bv.tum.de

M. Bader et al. (eds.), *Advanced Computing*, Lecture Notes in Computational Science and Engineering 93, DOI 10.1007/978-3-642-38762-3_1,

1 Introduction

Structural analysis with standard finite elements requires the discretization of the domain of interest into a finite element mesh, whose boundaries conform to the physical boundaries of the structure [3, 20]. While this constraint can be easily achieved for many applications in structural mechanics, it constitutes a severe bottleneck when highly complex geometries are concerned. An alternative pathway that avoids time-consuming mesh generation is provided by embedded domain methods, also known as immersed boundary methods [24–26, 29]. Their main idea consists of the extension of the physical domain of interest Ω_{phys} beyond its potentially complex boundaries into a larger embedding domain of simple geometry Ω, which can be meshed easily by a structured grid (see Fig. 1). The finite cell method (FCM) [12, 27, 39] is an embedded domain method, which combines the fictitious domain approach [6, 16, 31] with higher-order basis functions [45, 49], adaptive integration and weak imposition of unfitted Dirichlet boundary conditions [15, 37, 52]. To preserve consistency with the original problem, the influence of the fictitious domain extension Ω_{fict} is extinguished by penalizing its material parameters. For smooth problems of linear elasticity, the FCM has been shown to maintain exponential rates of convergence in the energy norm and thus allows for accurate structural analysis irrespective of the geometric complexity involved [32]. Moreover, it can be well combined with image-based geometric models typical for applications from biomechanics and material science [12, 36, 44]. Within the framework of the FCM for structural analysis, the following aspects have been examined so far: Topology optimization [28], thin-walled structures [33], local refinement strategies [40, 41, 43, 44], weak boundary conditions [36, 45], homogenization of porous and cellular materials [13], geometrically nonlinear problems [42, 45], and computational steering [22, 23, 50, 51].

The present contribution provides an overview of recent accomplishments in the finite cell method for structural mechanics. It is organized as follows: Sect. 2 provides a short introduction to the basic components of the finite cell method. Section 3 outlines the typical solution behavior for linear elastic problems and highlights important numerical properties. Section 4 shows the extension of the finite cell method to nonlinear elasticity. Section 5 outlines that FCM can be extended to problems beyond structural mechanics by the example of transport processes in porous media. Section 6 presents two application oriented numerical examples in three-dimensions, based on CAD and image-based geometric models. In Sect. 7, we conclude our presentation by a short summary and an outlook to future research.

2 A Brief Review of the Finite Cell Method

The following review provides a brief introduction to the main components, i.e. the fictitious domain concept, a higher-order approximation basis, adaptive integration

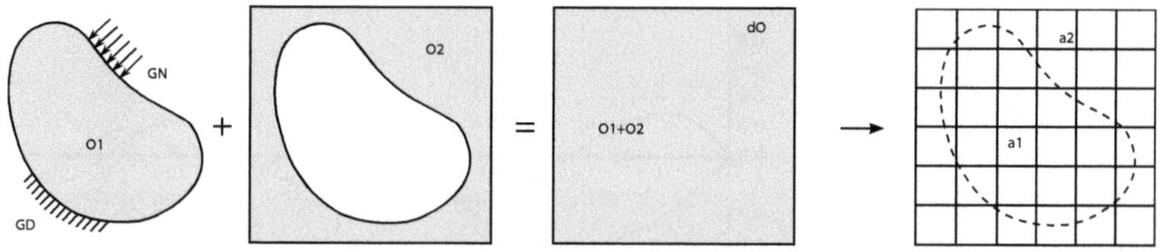

Fig. 1 The fictitious domain concept: The physical domain Ω_{phys} is extended by the fictitious domain Ω_{fict} into an embedding domain Ω to allow easy meshing of complex geometries. The influence of Ω_{fict} is penalized by α

and unfitted Dirichlet boundary conditions. We follow the presentation given in [45], focusing on the p- and the B-spline versions of the FCM.

2.1 *The Fictitious Domain Concept*

In the finite cell method, the domain to be analyzed is called the embedding domain Ω, which consists of the physical domain of interest Ω_{phys} and the fictitious domain extension Ω_{fict} as shown in Fig. 1. Analogous to standard finite element methods (FEM), the finite cell method for linear elastic problems is derived from the principle of virtual work

$$\delta W\left(\boldsymbol{u}, \delta \boldsymbol{u}\right) = \int_{\Omega} \boldsymbol{\sigma} : \left(\nabla_{sym}\, \delta \boldsymbol{u}\right)\, dV - \int_{\Omega_{phys}} \delta \boldsymbol{u} \cdot \boldsymbol{b}\, dV - \int_{\Gamma_N} \delta \boldsymbol{u} \cdot \boldsymbol{t}\, dA = 0 \tag{1}$$

where $\boldsymbol{\sigma}$, $\boldsymbol{b}$, $\boldsymbol{u}$, $\delta\boldsymbol{u}$ and ∇_{sym} denote the Cauchy stress tensor, body forces, displacement vector, test function and the symmetric part of the gradient, respectively [3,20]. Neumann boundary conditions are specified over the boundary of the embedding domain $\partial\Omega$, where tractions are zero by definition, and over Γ_N of the physical domain by traction vector $\boldsymbol{t}$ (see Fig. 1). The elasticity tensor $\boldsymbol{C}$ [3, 20] relating stresses and strains

$$\boldsymbol{\sigma} = \alpha \boldsymbol{C} : \boldsymbol{\varepsilon} \tag{2}$$

is complemented by a scalar factor α, which reads

$$\alpha\left(\boldsymbol{x}\right) = \begin{cases} 1.0 & \forall \boldsymbol{x} \in \Omega_{phys} \\ 10^{-q} & \forall \boldsymbol{x} \in \Omega_{fict} \end{cases} \tag{3}$$

penalizing the contribution of the fictitious domain. In Ω_{fict}, α must be chosen as small as possible, but large enough to prevent extreme ill-conditioning of the stiffness matrix [12,27]. Typical values of α range between 10^{-4} and 10^{-15}.

Using a structured grid of high-order elements (see Fig. 1), which will be called finite cells in the following, kinematic quantities are discretized as

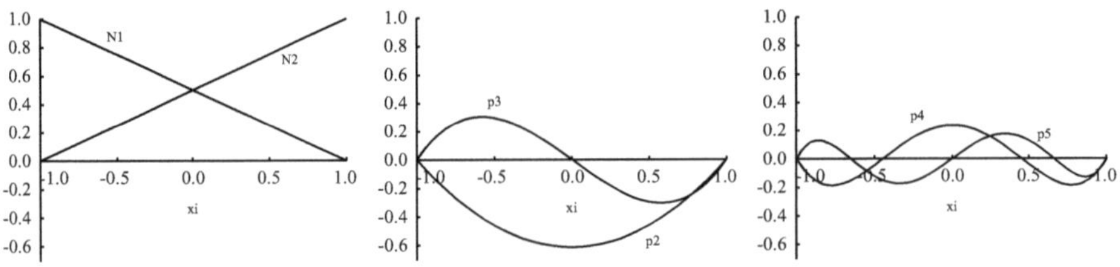

Fig. 2 Linear nodal modes N_j, $j=1,2$ and the first 4 integrated Legendre basis functions ϕ_j, $j=2,\ldots,5$ of the 1D p-version basis in the parameter space ξ

$$\boldsymbol{u} = \sum_{a=1}^{n} N_a \boldsymbol{u}_a \tag{4}$$

$$\delta\boldsymbol{u} = \sum_{a=1}^{n} N_a \delta\boldsymbol{u}_a \tag{5}$$

The sum of N_a denotes a finite set of n higher-order shape functions, and $\boldsymbol{u}_a$ and $\delta\boldsymbol{u}_a$ the corresponding vectors of unknown coefficients. Following the standard Galerkin approach [3, 20], inserting (4) and (5) into the weak form (1) produces a discrete finite cell representation

$$\boldsymbol{K}\boldsymbol{u} = \boldsymbol{f} \tag{6}$$

with stiffness matrix $\boldsymbol{K}$ and load vector $\boldsymbol{f}$. Due to the similarity to standard FEM, the implementation of FCM can make use of existing techniques.

2.2 *Higher-Order Approximation of Solution Fields*

The high-order basis originally applied in the FCM [12, 27] uses a regular mesh of elements of the *p-version of the FEM*, whose formulation is based on C^0 integrated Legendre polynomials [48, 49]. Corresponding basis functions in 1D are plotted in Fig. 2. The basis is hierarchical, so that an increase of the polynomial degree p by 1 is achieved by the addition of another function ϕ_j. Corresponding higher-dimensional bases can be constructed by tensor products of the 1D case. To limit the number of additional unknowns in 2D and 3D, the so-called *trunk space* is used instead of the full tensor product basis [48, 49].

The *B-spline version of the FCM* has been recently established as a suitable alternative [40, 41, 43, 45]. Its formulation is based on higher-order and smooth B-spline basis functions [30, 35], whose numerical advantages have been recently demonstrated in the context of isogeometric analysis [9, 21]. We use a single uniform B-spline patch, whose basis functions consist of uniform B-splines constructed from equidistant knots [30, 35] and can be interpreted as translated copies of each

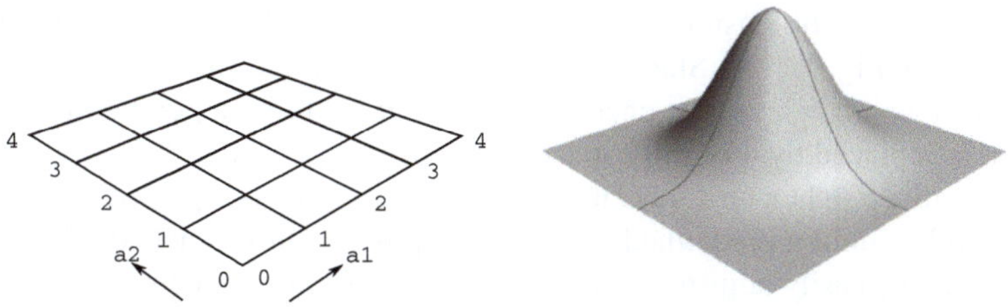

Fig. 3 Knot span cells in the parameter space $\{\xi, \eta\}$ (*left*) and corresponding bi-variate cubic B-spline (*right*)

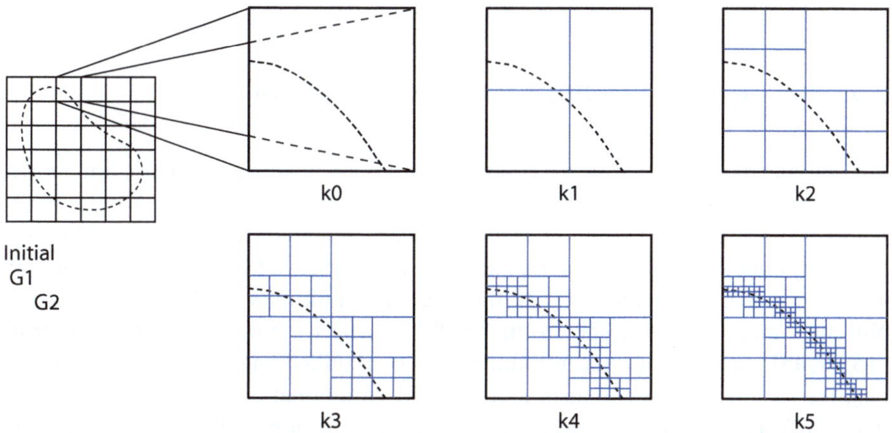

Fig. 4 2D sub-cell structure (*thin blue lines*) for adaptive integration of finite cells (*bold black lines*) that are cut by the geometric boundary (*dashed line*)

other [19]. Corresponding multivariate B-spline basis functions are obtained by taking the tensor product of the univariate components in each parametric direction. An example of a two-dimensional knot span structure and a corresponding bi-cubic uniform B-spline are shown in Fig. 3. Each knot span can be identified as a quadrilateral or hexahedral finite cell, respectively, with full Gaussian integration [40, 41]. The physical coordinates of the FCM grid can be generated from a simple linear transformation of the parametric space [45].

2.3 Adaptive Integration

The accuracy of numerical integration by Gauss quadrature [3, 20] is considerably influenced by discontinuities within cells introduced by the penalization parameter α of (3) [12, 27]. Therefore, the FCM uses composed Gauss quadrature in cells cut by geometric boundaries, based on a hierarchical decomposition of the original cells

[12, 44, 45]. In two dimensions, the sub-cell structure can be built up in the sense of a quadtree (see Fig. 4) [38]. Starting from the original finite cell of level $k = 0$, each sub-cell of level $k = i$ is first checked whether it is cut by a geometric boundary. If true, it is replaced by 4 equally spaced cells of level $k = i + 1$, each of which is equipped with $(p+1) \times (p+1)$ Gauss points. Partitioning is repeated for all cells of current level k, until a predefined maximum depth $k = m$ is reached. The quadtree approach can be easily adjusted to 1D or 3D by binary trees or octrees, respectively. It is easy to implement and keeps the regular grid structure of the FCM. To clearly distinguish between finite cell and sub-cell meshes, finite cells are plotted in black and integration sub-cells are plotted in blue lines throughout this paper (see Fig. 4).

2.4 Imposition of Unfitted Boundary Conditions

For complex domains, boundary conditions are defined along geometric boundaries cutting arbitrarily through finite cells. Neumann boundary conditions can be incorporated by simple integration over the Neumann boundary Γ_N (see (1)). Dirichlet boundary conditions require an imposition in a weak sense by variational techniques such as the penalty method [1, 52], the Lagrange multiplier method [15, 16, 53] or Nitsche's method [4, 14, 17]. In the FCM, Nitsche's method is usually preferred [36, 37, 45], since it does not introduce additional unknowns, leads to a symmetric, positive definite stiffness matrix and satisfies variational consistency in the sense that solutions of the weak form can be shown to be solutions of the original boundary value problem.

From a practical point of view, the integration over unfitted boundaries is accomplished by introducing a triangular mesh of the boundary surfaces. Generating a triangulation of a 3D surface is a standard task, for which a variety of efficient algorithms and tools are available. In particular, it is orders of magnitude less complex and less expensive than the generation of a full volumetric discretization of a complex 3D object. Corresponding mesh generation in the framework of the FCM for a boundary representation of solids and for voxel-based data obtained from CT scans is addressed in detail in [12].

3 Basic Numerical Properties of the FCM

For the illustration of the typical solution behavior, a linear elastic uni-axial rod is examined, for which geometry, material and boundary conditions are specified in Fig. 5. Its middle part represents the fictitious domain Ω_{fict}, whose Young's modulus E is penalized with parameter $\alpha = 10^{-8}$. The example approximates the situation of two separate rods. The right one undergoes a rigid body movement Δu and the left one is subjected to a sine load $f_{\sin}$. The FCM discretizations considered consist of 2 p-version finite cells and 11 knot span cells as shown in Fig. 5. Due to the different

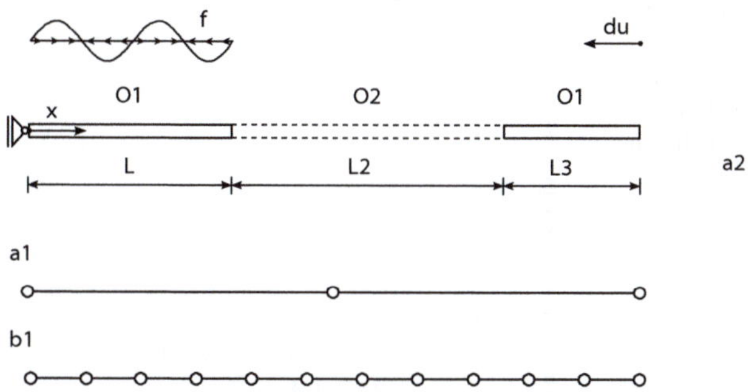

Fig. 5 Uni-axial rod with geometric boundaries at $X = 1.0$ and $X = 2\frac{1}{3}$

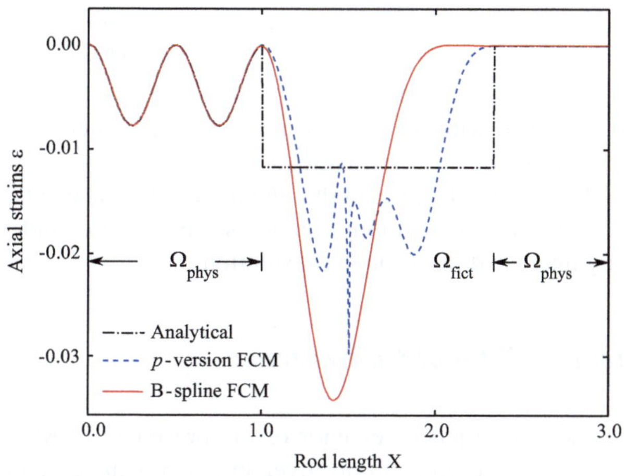

Fig. 6 Smooth extension of the FCM solutions ($p = 15$) vs. discontinuous analytical solution ($\alpha = 10^{-8}$) for the linear elastic strains of the 1D example

construction of the bases, the B-spline version requires a denser knot span grid than the p-version in order to achieve a comparable amount of degrees of freedom (dofs). For all computations of this section, adaptive sub-cells of depth $k = 20$ are used to minimize the integration error in cells cut by geometric boundaries.

3.1 Smooth Extension of Solution Fields

The p- and B-spline versions of the FCM produce solution fields, which extend smoothly into the fictitious domain despite the discontinuities of the analytical solution. This is illustrated in Fig. 6, which compares the analytical strains to the numerical strains of the p- and B-spline versions. The importance of the smooth extension of the FCM solution for the overall convergence behavior of the finite cell

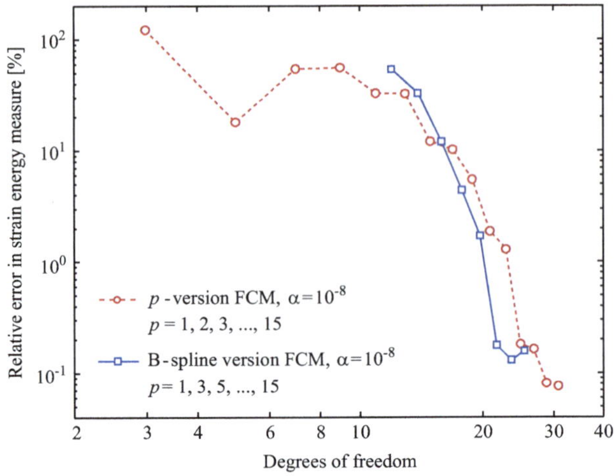

Fig. 7 Convergence in strain energy obtained by p-refinement in the given p-version and B-spline discretizations of the linear elastic 1D example

method can be explained with the help of the penalty parameter α in conjunction with the best approximation property to the total strain energy [47] (see for example [43, 45] for details). It should be noted that the large difference between analytical and FCM solution fields in the fictitious domain (see Fig. 6) is completely irrelevant, since we are only interested in the physical solution.

3.2 *Exponential Rates of Convergence*

Figure 7 shows the convergence behavior of the presented FCM schemes, if the polynomial degree of the discretizations given in Fig. 5 is increased from $p = 1$ to 15. Both the p-version and the B-spline version of the FCM converge exponentially. The penalization parameter $\alpha = 10^{-8}$ as well as the integration error in cut cells lead to a flattening of the convergence curves. The present example shows that the p- and B-spline bases exhibit an equivalent solution behavior within the FCM and achieve a comparable performance in terms of error level, rates of convergence and flattening of the convergence curve, although their high-order approximation bases are very different. Further numerical benchmarks in higher dimensions can be found in [43, 45] that show optimal rates of convergence under h-refinement, the stability and accuracy of weak boundary conditions and the competitive quality of the solution and its derivatives along the geometric boundaries in cut cells with respect to standard body-fitted finite element methods.

4 Extension to Nonlinear Elasticity

The finite cell method can be extended to geometrically nonlinear elasticity on the basis of the logarithmic strain measure [7] and the Hencky hyperelastic material model [10]. An extensive review of the mathematical model and the pertinent continuum mechanics in the framework of the FCM can be found in [45]. In the present scope, we focus on our 1D rod example of Fig. 5, for which the corresponding geometrically nonlinear formulation simplifies to

$$\Psi = \alpha \frac{E}{2} (\ln \lambda)^2 \tag{7}$$

$$\sigma = \alpha \frac{E}{J} \ln \lambda \tag{8}$$

$$c = \alpha \frac{E}{J} - 2\sigma \tag{9}$$

with axial stretch λ, the determinant of the deformation gradient $J = \lambda^{1-2\nu}$ and the strain energy function Ψ [7]. To illustrate the influence of large deformations within the fictitious domain, the sine load f_{sin} of Fig. 5 is neglected for now and the prescribed displacement is set to a large value of $\Delta u = 1.0$. The physical stresses should be zero, since a rigid body movement of the right part of the rod is approximated. The exact stress solution, which can be derived analytically according to [40], is plotted in Fig. 8a for 10 displacement load increments between 0 and Δu and $\alpha = 10^{-5}$. In the following, all numerical examples are computed with the B-spline version of the FCM, but equivalent results can be derived for the p-version (see [45]).

4.1 The Standard FCM Formulation

The standard FCM is based on the application of the same geometrically nonlinear formulation over the complete embedding domain Ω. However, numerical experiments reveal that for α smaller than 10^{-5}, the determinant of the deformation gradient falls below zero at some integration point within Ω_{fict}, which inevitably terminates the computation. With α as large as 10^{-5}, the penalization of (3) is unable to sufficiently eliminate the influence of Ω_{fict}, so that a considerable modeling error is introduced. In addition, nonlinear strains increasingly outweigh the penalization by α, since they are able to grow without bounds [45]. The corresponding stress solution obtained with 16 knot span cells in the sense of Fig. 5 is plotted in Fig. 8b. It exhibits large oscillations throughout the discontinuous cells and the corresponding convergence deteriorates to a low algebraic rate (see Fig. 9). The standard FCM formulation thus suffers from a conflict of interest between stable analysis (increase of α) on the one hand and a reduction of the contribution of Ω_{fict} (decrease of α) on the other.

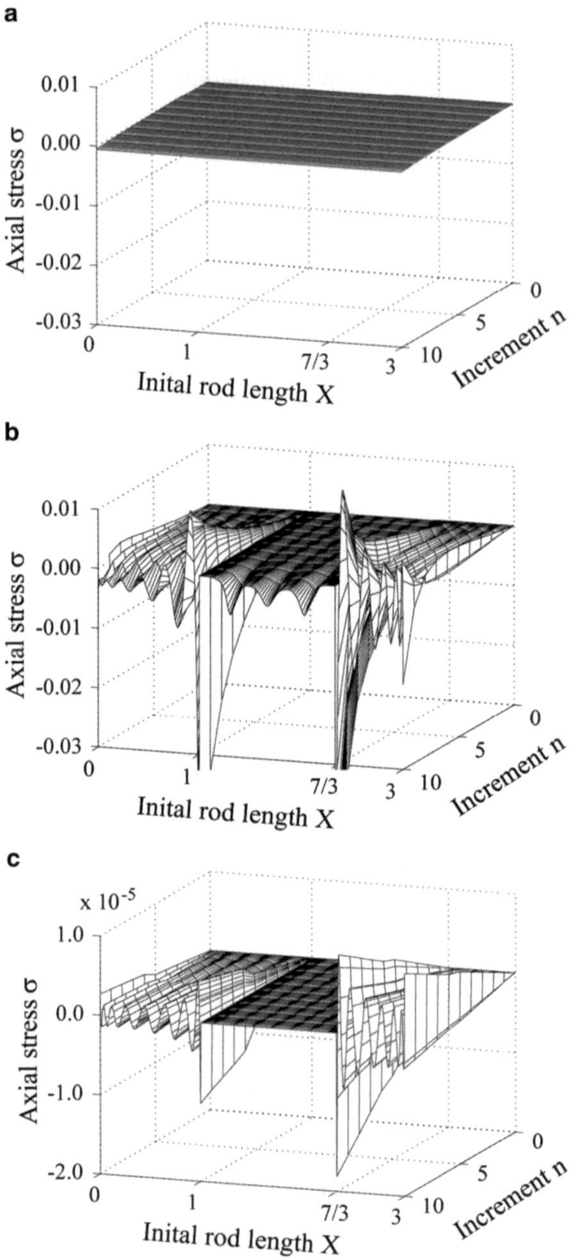

Fig. 8 The stress solution of the geometrically nonlinear rod is obtained without and with deformation resetting (16 knot span elements of $p = 15$). Note that deformation resetting reduces the stress oscillations by three orders of magnitude. (**a**) Analytical stress solution. (**b**) Computed with the standard B-spline version of the FCM. (**c**) Computed with the B-spline version of the FCM and deformation resetting

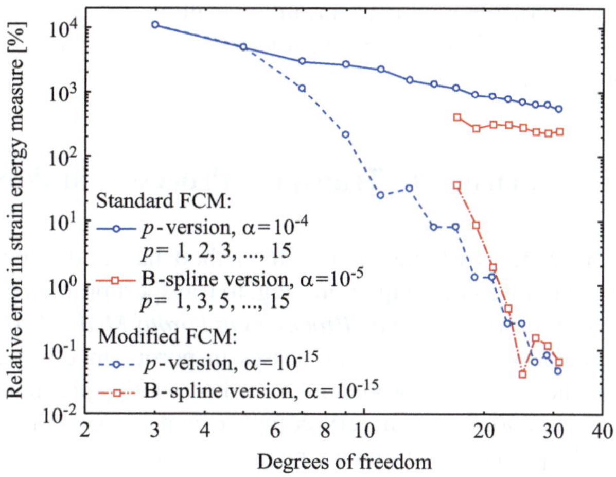

Fig. 9 Convergence without and with deformation resetting for the nonlinear rod

4.2 A Modified Formulation Based on Deformation Resetting

Numerical experiments reveal that problems with the uniqueness of the deformation map occur at the location of maximum deformation within the fictitious domain Ω_{fict}. This motivates the following simple manipulation after each Newton iteration i

$$\varphi^i(\mathbf{X}) = \begin{cases} \boldsymbol{x}^i & \text{deformed configuration } \forall \boldsymbol{X} \in \Omega_{phys} \\ \boldsymbol{X} & \text{reset to reference configuration } \forall \boldsymbol{X} \in \Omega_{fict} \end{cases} \tag{10}$$

where $\boldsymbol{\varphi}^i$ and $\boldsymbol{x}^i$ denote the deformation map and the deformed configuration after the ith Newton step. According to (10), the deformation is repeatedly reset to the initial undeformed state to erase the complete deformation history within the fictitious domain Ω_{fict}. This does not affect the physical consistency and accuracy of the solution in the physical domain Ω_{phys}, provided that the influence of Ω_{fict} is mitigated by a sufficiently strong penalization. Furthermore, the assumption of (10) supersedes the calculation of the deformation gradient [45], so that any stability issues resulting from the numerical computation of the deformation gradient are automatically avoided. The corresponding stress solution is plotted in Fig. 8c, where the oscillatory behavior of Fig. 8b is considerably reduced by several orders of magnitude. Moreover, the deformation resetting can be efficiently implemented by exploiting the coincidence of linear and geometrically nonlinear elasticity at the deformation and stress-free reference configuration [45].

To test convergence in an energy measure, the uni-axial rod of Fig. 5 is considered with sine-load $f_{\sin}$ and $\Delta u = 1.0$. The convergence under p-refinement is plotted in Fig. 9 for the p- and B-spline versions. For the standard FCM formulation, it illustrates the convergence decay to a low algebraic rate as a consequence of the insufficient penalization in conjunction with oscillatory stresses. The modified geometrically nonlinear formulation allows for a decrease of the penalty parameter to $\alpha = 10^{-15}$, which restores the ability of the FCM schemes to achieve exponential convergence.

5 Extension to Transport Processes in Porous Media

The concept of the finite cell method has also been applied to other types of partial differential equations, e.g. to transport problems within the MAC project B5 *Transport and Reaction Processes in Porous Media*. From a geometric point of view, simulation of transport processes in porous media may be similarly challenging as the structural problems described in the previous sections. The domain of computation is often very complex, and the generation of a mesh that conforms to the porous structure may be highly involved.

We start from the weak formulation of the transport equation that reads

$$\int_{\Omega} [\mathbf{q}c \cdot \nabla w + (\theta \nu)\nabla c \cdot \nabla w] d\Omega = \int_{\Omega} w f d\Omega \tag{11}$$

where c is the concentration, w is a test function and f is a source term. The quantities $\mathbf{q}$ and $(\theta\nu)$ denote the Darcy's velocity and the effective diffusion coefficient, respectively [18]. In the sense of Fig. 1, the porous flow domain Ω_{phys} is embedded in a larger domain, which is meshed by a simple Cartesian grid. The bilinear form associated to the weak form of (11) is extended to the fictitious domain, and the quantities $\mathbf{q}$ and $(\theta\nu)$ are multiplied by a penalty factor α in the sense of (3):

$$\begin{cases} \mathbf{q}_e = \alpha \cdot \mathbf{q} \\ (\theta \nu)_e = \alpha \cdot \theta \nu \end{cases} \tag{12}$$

Quadtree-based numerical integration is performed for cells cut by the boundary of the flow domain (see Sect. 2.3) and a Bubnov-Galerkin Ansatz is made with high-order basis functions.[1]

Figure 10 shows results of a test case for the simulation of transport through porous media, where for simplicity the velocity $\mathbf{q}$ of the transporting fluid was assumed to be constant along the x-axis throughout the domain. Although this can only be assumed for a limited range of physical problems, it does not impose a restriction with respect to the validation of the approach, as long as the transport velocity $\mathbf{q}$ is taken as an upper bound of the expected true flow velocity. The square domain is discretized by 8×8 finite cells of polynomial degree $p = 8$. Figure 10b illustrates the profile of concentration c along the diagonal cut line. A reference

[1] It is worthwhile to note that high-order basis functions are significantly more stable than low-order functions for flow problems moderately dominated by convection [8].

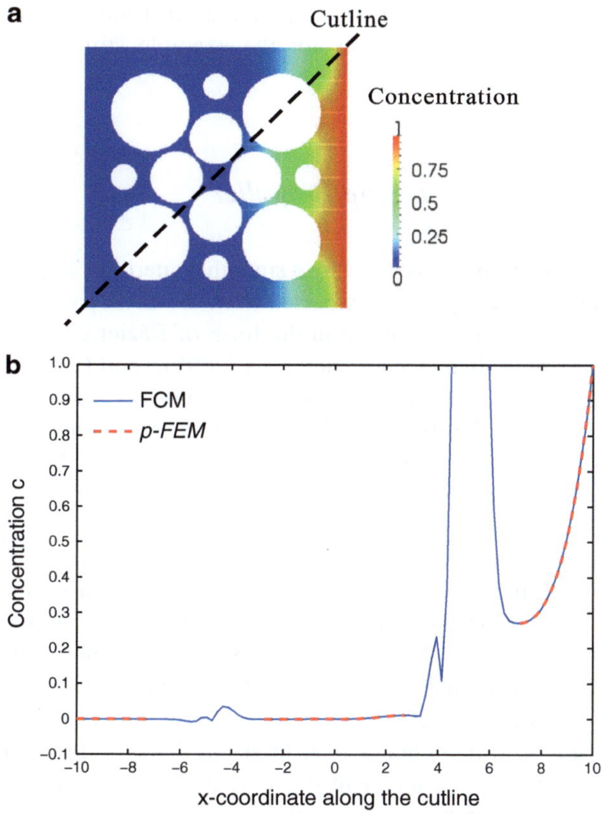

Fig. 10 FCM analysis of porous media: The test case consists of a square domain with impermeable spheres. Dirichlet constraints of $c = 0$ and $c = 1$ are imposed on the left and right boundaries, respectively, and no flow conditions on the upper and lower boundaries. (**a**) Concentration profile for a Péclet number of $Pe = 1$ [11]. (**b**) Finite cell solution along the *diagonal cut line* as compared to a body-fitted p-FEM reference solution

solution obtained by a refined computation of a boundary fitted mesh is compared to the FCM solution for a Péclet number of $Pe = 1$ [11, 18]. Similar to structural problems, the FCM solution shows very good quality, even in areas, where only "narrow" flow bridges between obstacles are present.

6 Application Oriented Examples: Structural Analysis of CAD and Image-Based Geometric Models

In the following, we illustrate the benefits of the finite cell method in terms of simple mesh generation for very complex geometries by two application oriented examples, which are described by a CAD (computer aided design) based T-spline surface and a CT (computed tomography) based voxel model, respectively. A more detailed description and further computations for the example structures can be found in [43–45].

6.1 Modal Analysis of a Ship Propeller

The geometry of the propeller is given by a smooth, watertight T-spline surface (i.e., there are no gaps or overlaps). It is exported from the CAD package Rhino [34] in conjunction with the T-spline plug-in in the form of Bézier elements as shown in Fig. 11a. Its maximum diameter and height is 0.695 m and 0.334 m, respectively, and it is made out of steel with Young's modulus $2.1 \cdot 10^{11}$ N/m^2, Poisson's ratio 0.28 and density 7,850 kg/m^3. The structure can neither be characterized as a typical shell nor as a true solid. Configurations like this usually require specialized and time consuming meshing procedures to produce good quality discretizations.

The finite cell method circumvents the meshing challenge completely, which we demonstrate in the following with the B-spline version of the FCM. First, the complete structure is embedded in a regular grid of axis-aligned B-splines of polynomial degree $p = 3$ (see Fig. 11b). Second, all knot span cells without support in the propeller domain are eliminated from the discretization (see Fig. 12a). The decision whether an element is to be kept or not is based on a simple point location query, which checks if at least one integration point is located in Ω_{phys}. It can be efficiently implemented for example by search algorithms based on special space-partitioning data structures such as k-d trees [5, 38]. An axis-aligned discretization with elements of the same size does not account for the inhomogeneous thickness of the different regions of the structure. In a third step, we therefore apply two levels of hierarchical refinement to the propeller blades, while we leave the discretization of the central hub as it is, to achieve a homogeneous resolution of the two different thicknesses [43]. In a fourth step, we equip each element cut by the geometric boundary by additional sub-cells, which are organized in an octree of depth two (see Fig. 13). Each sub-cell contains $4 \times 4 \times 4$ Gauss points, leading to an aggregation of integration points in cut elements to accurately take into account the geometric boundary during numerical integration. The contribution to stiffness and mass matrices that result from integration points located outside the propeller domain Ω_{phys} are penalized by factor $\alpha = 10^{-3}$. The hierarchically refined mesh of Fig. 12b is analysis suitable and is used in combination with the sub-cells of Fig. 13 to conduct a modal analysis of the structure, where the mass matrix is lumped according to the row sum method [20]. Figure 14 illustrates the first mode shapes.

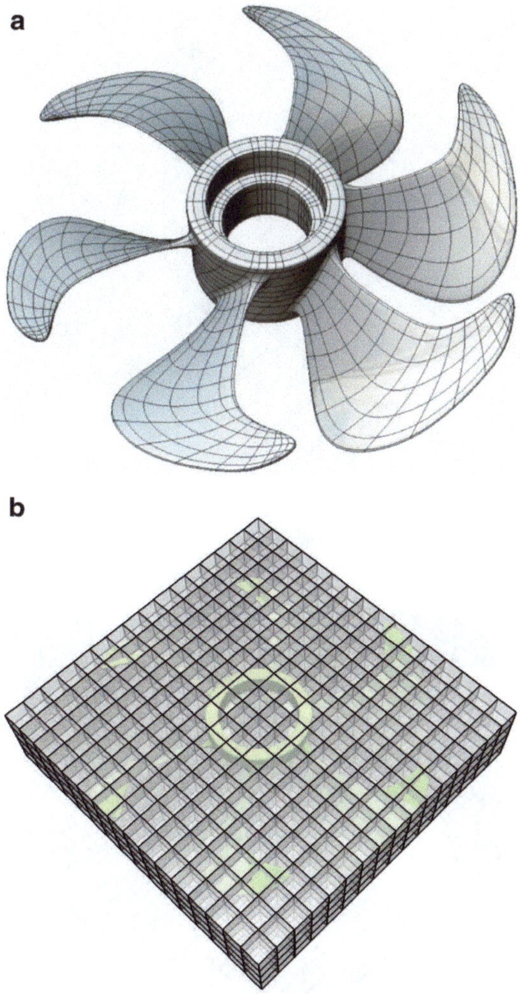

Fig. 11 Ship propeller example: CAD based geometry description and finite cell discretization. (**a**) Bézier elements of a T-spline surface (Output from CAD package Rhino with T-spline plug-in). (**b**) The complete structure is immersed in a bounding box of $16 \times 16 \times 4$ axis-aligned cubic B-spline elements

6.2 Large Deformation Analysis of an Open-Cell Aluminium Foam

Metal foams provide high stiffness at reduced weights, and are therefore frequently used for lightweight structures in automotive and aerospace applications [2]. The p-version of the FCM is applied to simulate a compression test for an aluminium foam sample of size $20 \times 20 \times 20$ mm, discretized by a structured grid of $5 \times 5 \times 5$

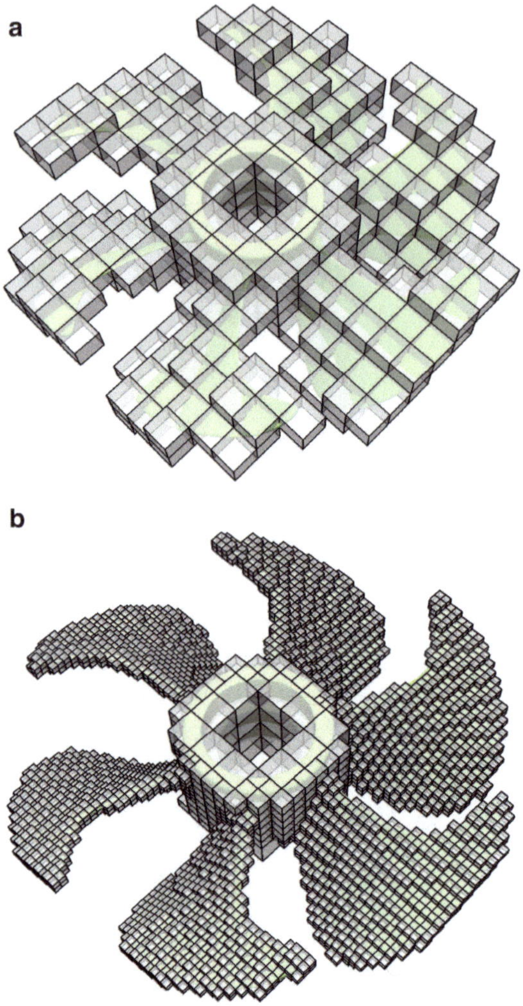

Fig. 12 Ship propeller example: The role of hierarchical refinement. (**a**) Deletion of elements without support in the propeller domain creates a reduced set of elements, which homogeneously resolve the structure irrespective of the local thickness. (**b**) Hierarchical refinement of the propeller blades achieves a homogeneous through-the-thickness resolution

high-order finite cells. Its internal geometry is provided by voxels with a resolution of $1,024$ in each Cartesian direction, each of which encodes α. Figure 15a, b show the complete voxel model of the sample cube and the physical voxels of material index 1 associated with aluminium, respectively, in a coarsened resolution of 128^3. The foam sample is assumed as part of a larger specimen, which is uniformly compressed along the vertical axis. Corresponding boundary conditions are specified as follows [46]: Displacements normal to the top surface are gradually increased to 1.6 mm (8 % compressive deformation), modelling the influence of

Fig. 13 Sub-cell partitioning of elements cut by the geometric boundary. The adaptive decomposition scheme shown in Fig. 4 is carried out up to level $k = 2$

Fig. 14 The first, second and third mode shapes exhibit a rotational symmetry around the center, corresponding to the three pairs of opposing propeller blades. We display mode 2

a testing machine, whereas the displacements normal to all other surfaces are fixed due to the bottom support and the influence of the surrounding material of the specimen. The aluminium foam is characterized by Young's modulus $E = 70.000\,\text{N/mm}^2$, penalized by $\alpha = 10^{-12}$ at all integration points in Ω_{fict}, and Poisson's ratio $\nu = 0.35$.

The finite cell method is able to directly operate on the voxel model, which provides a basis for a simple point location query (see [45]). In particular, we avoid the costly transformation of the voxel model into a surface model by image-based software, which is required as a basis for mesh generation by standard body-fitted

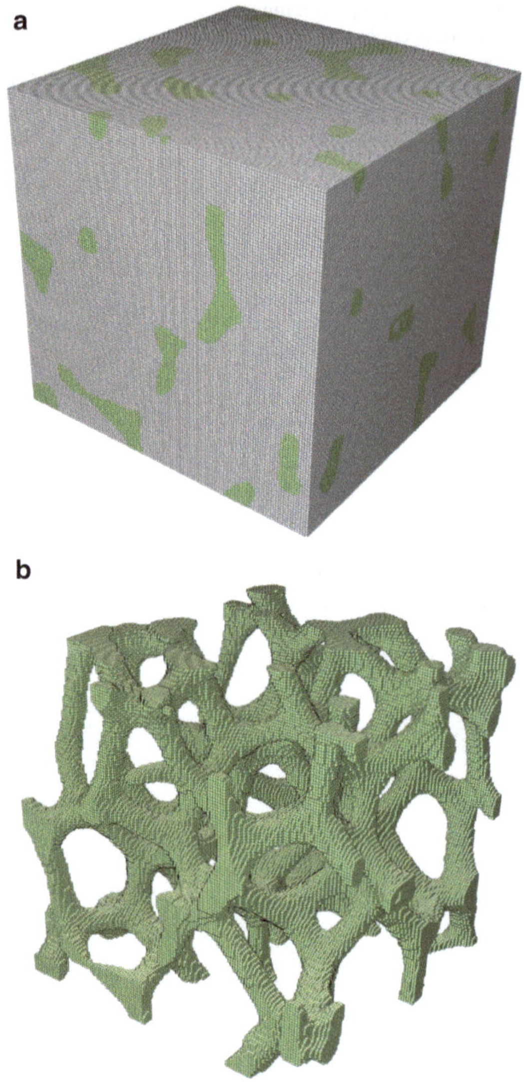

Fig. 15 CT-based voxel model of an aluminium foam sample. For better visibility, the original resolution of $1{,}024^3$ voxels is reduced to 128^3. (**a**) Voxelized sample cube. (**b**) Voxels with $b_{vox} = 1$

simulation methods. For the present p-version mesh of polynomial degree $p = 7$ with 3 levels of sub-cells (21,492 dofs; 24,947 sub-cells; approx. 12.75 million Gauss points), analysis of the foam could be accomplished by our in-house FCM code in about 4 h.[2] Since the major cost of FCM results from the large number of sub-cells with full Gauss integration, a major performance gain is achieved by the shared memory parallelization of the loop that computes local stiffness

[2]Using eight threads on two interconnected Intel(R) Xeon(R) W5590 @ 3.33 GHz.

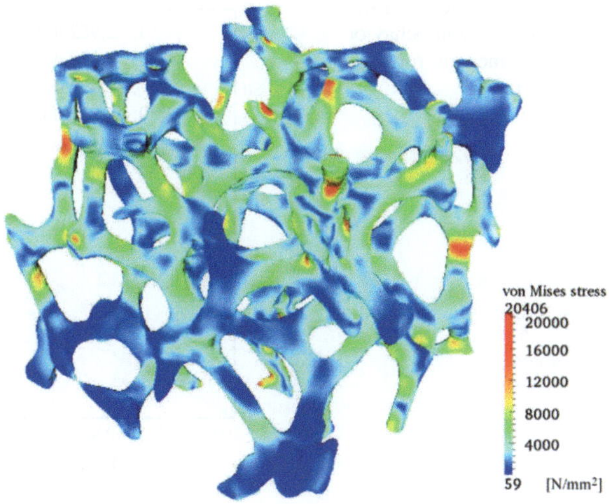

Fig. 16 Von Mises stress of the metal foam sample plotted on the deformed configuration. The results are obtained from a discretizations with $5 \times 5 \times 5$ finite cells of polynomial degree $p = 7$

matrices for cells and sub-cells with subsequent assembly into the global system matrix. A *parallel for* construct creates a team of n threads to execute the main loop over sub-cells in parallel, where n is the number of threads available. With $n = 8$, we achieved a strong speed up of the loop of around 5. In addition, we would like to mention here that a very fast variant of the FCM has been developed (see [22, 23, 50, 51]), relying on pre-integration of sub-element matrices. This implementation is up to three orders of magnitude faster than the FCM based on standard quadrature described herein, and thus allows even for real-time simulations of complex 3D structures.

The resulting von Mises stresses shown in Fig. 16 exhibit accurate localization of stress concentrations at the convex sides of the foam members, which agrees well with engineering experience. Figure 17 plots the equivalent force obtained from integration of the normal stress over the top surface vs. the prescribed displacement of the top surface for different polynomial degrees p. It can be observed that the increase of p improves the reproduction of the geometrically nonlinear behavior of the foam.

7 Summary and Outlook

The present contribution provided a review of recent developments of the finite cell method (FCM) for the analysis of complex structures. We briefly summarized the basic components of the FCM technology, i.e. the fictitious domain concept, high-order basis functions, adaptive integration and weak boundary conditions, and outlined its basic numerical properties for linear elasticity, i.e. the smooth extension of solution fields beyond the physical domain as well as exponential rates of convergence in energy norm. We then summarized the concept of deformation resetting, which enables the extension of the FCM to nonlinear elasticity. Finally,

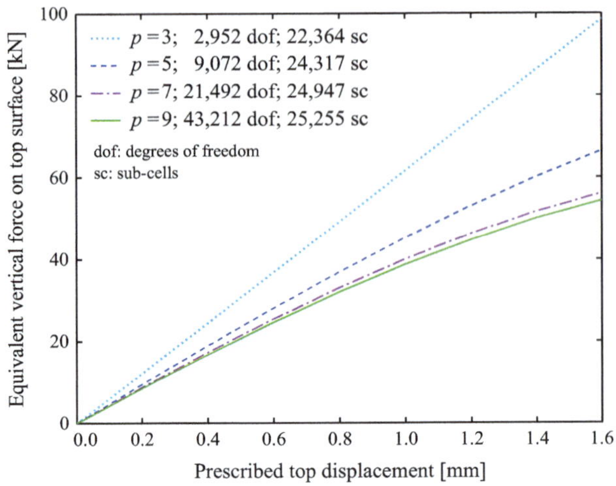

Fig. 17 Convergence of the force-displacement behavior under p-refinement for the foam sample

we illustrated with two application examples, i.e. a ship propeller and a metal foam sample, that the benefits of the finite cell method in terms of almost no mesh generation for complex structures can be achieved for both CAD based and explicit image-based geometric models.

Based on these results, we believe that the finite cell method has great potential for the accurate analysis of very complex structures, and a plethora of very promising aspects are still open, such as the analysis of topology changes and moving boundaries, for which embedded domain methods such as the FCM offer significant advantages over ALE-type approaches, or the introduction of FCM suitable coupling schemes for multiphysics problems, which stand at the forefront of today's challenges in computational science.

Acknowledgements D. Schillinger, Q. Cai and R.-P. Mundani gratefully acknowledge support from the Munich Centre of Advanced Computing (MAC) and the International Graduate School of Science and Engineering (IGSSE) at the Technische Universität München. D. Schillinger gratefully acknowledges support from the German National Science Foundation (Deutsche Forschungsgemeinschaft DFG) under grant number SCHI 1249/1-1. The authors thank T.J.R. Hughes, M. Ruess and M.A. Scott for their help with the analysis of the ship propeller.

References

1. Babuška, I.: The finite element method with penalty. Math. Comput. **27**(122), 221–228 (1972)
2. Banhart, J.: Manufacture, characterization and application of cellular metals and metal foams. Prog. Mater. Sci. **46**, 559–632 (2001)
3. Bathe, K.J.: Finite Element Procedures. Prentice-Hall, Englewood Cliffs (1996)
4. Bazilevs, Y., Hughes, T.: Weak imposition of dirichlet boundary conditions in fluid mechanics. Comput. Fluids **36**, 12–26 (2007)

5. Bindick, S., Stiebler, M., Krafczyk, M.: Fast kd-tree-based hierarchical radiosity for radiative heat transport problem. Int. J. Numer. Methods Eng. **86**, 1082–1100 (2009)
6. Bishop, J.: Rapid stress analysis of geometrically complex domains using implicit meshing. Comput. Mech. **30**, 460–478 (2003)
7. Bonet, J., Wood, R.: Nonlinear Continuum Mechanics for Finite Element Analysis. Cambridge University Press, Cambridge (2008)
8. Cai, Q.: The finite cell method for transport and reaction processes in porous media. Ph.D. thesis, Technische Universität München (2013)
9. Cottrell, J., Hughes, T., Bazilevs, Y.: Isogeometric Analysis: Towards Integration of CAD and FEA. Wiley, Chichester/Hoboken (2009)
10. de Souza Neto, E., Perić, D., Owen, D.: Computational Methods for Plasticity: Theory and Applications. Wiley, Chichester (2008)
11. Donea, J., Huerta, A.: Finite Element Methods for Flow Problems. Wiley, Chichester/Hoboken (2003)
12. Düster, A., Parvizian, J., Yang, Z., Rank, E.: The finite cell method for threedimensional problems of solid mechanics. Comput. Methods Appl. Mech. Eng. **197**, 3768–3782 (2010)
13. Düster, A., Sehlhorst, H.G., Rank, E.: Numerical homogenization of heterogeneous and cellular materials utilizing the finite cell method. Comput. Mech. **50**, 413–431 (2012)
14. Embar, A., Dolbow, J., Harari, I.: Imposing dirichlet boundary conditions with Nitsche's method and spline-based finite elements. Int. J. Numer. Methods Eng. **83**, 877–898 (2010)
15. Fernández-Méndez, S., Huerta, A.: Imposing essential boundary conditions in mesh-free methods. Comput. Methods Appl. Mech. Eng. **193**, 1257–1275 (2004)
16. Glowinski, R., Kuznetsov, Y.: Distributed lagrange multipliers based on fictitious domain method for second order elliptic problems. Comput. Methods Appl. Mech. Eng. **196**, 1498–1506 (2007)
17. Hansbo, A., Hansbo, P.: An unfitted finite element method, based on Nitsche's method, for elliptic interface problems. Comput. Methods Appl. Mech. Eng. **191**, 537–552 (2002)
18. Helmig, R.: Multiphase Flow and Transport Processes in the Subsurface: A Contribution to the Modeling of Hydrosystems. Springer, Berlin/New York (1997)
19. Höllig, K.: Finite Element Methods with B-Splines. Society for Industrial and Applied Mathematics, Philadelphia (2003)
20. Hughes, T.: The Finite Element Method: Linear Static and Dynamic Finite Element Analysis. Dover Publications, Mineola (2000)
21. Hughes, T., Cottrell, J., Bazilevs, Y.: Isogeometric analysis: CAD, finite elements, NURBS, exact geometry and mesh refinement. Comput. Methods Appl. Mech. Eng. **194**, 4135–4195 (2005)
22. Knežević, J., Frisch, J., Mundani, R.P., Rank, E.: Interactive computing framework for engineering applications. J. Comput. Sci. **7**, 591–599 (2011)
23. Knežević, J., Mundani, R.P., Rank, E.: Interactive computing – virtual planning of hip-joint surgeries with real-time structure simulations. Int. J. Model. Optim. **1**(4), 308–313 (2011)
24. Löhner, R., Cebral, R., Camelli, F., Appanaboyina, S., Baum, J., Mestreau, E., Soto, O.: Adaptive embedded and immersed unstructured grid techniques. Comput. Methods Appl. Mech. Eng. **197**, 2173–2197 (2008)
25. Mittal, R., Iaccarino, G.: Immersed boundary methods. Annu. Rev. Fluid Mech. **37**, 239–261 (2005)
26. Neittaanmäki, P., Tiba, D.: An embedding domains approach in free boundary problems and optimal design. SIAM J. Control Optim. **33**(5), 1587–1602 (1995)
27. Parvizian, J., Düster, A., Rank, E.: Finite cell method: *h*- and *p*- extension for embedded domain methods in solid mechanics. Comput. Mech. **41**, 122–133 (2007)
28. Parvizian, J., Düster, A., Rank, E.: Toplogy optimization using the finite cell method. Comput. Mech. **13**, 57–78 (2012)
29. Peskin, C.: The immersed boundary method. Acta Numer. **11**, 479–517 (2002)
30. Piegl, L., Tiller, W.: The NURBS Book. Springer, Berlin/New York (1997)

31. Ramière, I., Angot, P., Belliard, M.: A fictitious domain approach with spread interface for elliptic problems with general boundary conditions. Comput. Methods Appl. Mech. Eng. **196**, 766–781 (2007)
32. Rank, E., Düster, A., Schillinger, D., Yang, Z.: The finite cell method: high order simulation of complex structures without meshing. In: Computational Structural Engineering, pp. 87–92. Springer, Dordrecht/New York (2009)
33. Rank, E., Ruess, M., Kollmannsberger, S., Schillinger, D., Düster, A.: Geometric modeling, isogeometric analysis and the finite cell method. Comput. Methods Appl. Mech. Eng. **249–250**, 104–115 (2012)
34. Rhino3d.: Rhinoceros – NURBS modeling for Windows (2012). http://www.rhino3d.com
35. Rogers, D.: An Introduction to NURBS with Historical Perspective. Morgan Kaufmann, San Francisco (2001)
36. Ruess, M., Tal, D., Trabelsi, N., Yosibash, Z., Rank, E.: The finite cell method for bone simulations: verification and validation. Biomech. Model. Mechanobiol. **11**(3), 425–437 (2012)
37. Ruess, M., Schillinger, D., Bazilevs, Y., Varduhn, V., Rank, E.: Weakly enforced essential boundary conditions for NURBS-embedded and trimmed NURBS geometries on the basis of the finite cell method. Int. J. Numer. Methods Eng. **95**(10), 811–846 (2013)
38. Samet, H.: Foundations of Multidimensional and Metric Data Structures. Morgan Kaufmann, Amsterdam/Boston (2006)
39. Schillinger, D.: The *p*- and B-spline versions of the geometrically nonlinear finite cell method and hierarchical refinement strategies for adaptive isogeometric and embedded domain analysis. Ph.D. thesis, Technische Universität München (2012)
40. Schillinger, D., Kollmannsberger, S., Mundani, R.P., Rank, E.: The finite cell method for geometrically nonlinear problems of solid mechanics. IOP Conf. Ser.: Mater. Sci. Eng. **10**, 012170 (2010)
41. Schillinger, D., Rank, E.: An unfitted *hp* adaptive finite element method based on hierarchical B-splines for interface problems of complex geometry. Comput. Methods Appl. Mech. Eng. **200**(47–48), 3358–3380 (2011)
42. Schillinger, D., Ruess, M., Düster, A., Rank, E.: The finite cell method for large deformation analysis. In: Proceedings in Applied Mathematics and Mechanics, vol. 11, pp. 271–272 (2011)
43. Schillinger, D., Dede', L., Scott, M., Evans, J., Borden, M., Rank, E., Hughes, T.: An isogeometric design-through-analysis methodology based on adaptive hierarchical refinement of NURBS, immersed boundary methods, and T-spline CAD surfaces. Comput. Methods Appl. Mech. Eng. **249–250**, 116–150 (2012)
44. Schillinger, D., Düster, A., Rank, E.: The *hp-d* adaptive finite cell method for geometrically nonlinear problems of solid mechanics. Int. J. Numer. Methods Eng. **89**, 1171–1202 (2012)
45. Schillinger, D., Ruess, M., Zander, N., Bazilevs, Y., Düster, A., Rank, E.: Small and large deformation analysis with the *p*- and B-spline versions of the finite cell method. Comput. Mech. **50**(4), 445–478 (2012)
46. Sehlhorst, H.G., Jänicke, J., Düster, A., Rank, E., Steeb, H., Diebels, S.: Numerical investigations of foam-like materials by nested high-order finite element methods. Comput. Mech. **45**, 45–59 (2009)
47. Strang, G., Fix, G.: An Analysis of the Finite Element Method. Prentice-Hall, Englewood Cliffs (1973)
48. Szabó, B., Babuška, I.: Finite Element Analysis. Wiley, New York (1991)
49. Szabó, B., Düster, A., Rank, E.: The *p*-version of the finite element method. In: Stein, E., de Borst, R., Hughes, T. (eds.) Encyclopedia of Computational Mechanics, vol. 1, chap. 5, pp. 119–139. Wiley, Chichester (2004)
50. Yang, Z., Kollmannsberger, S., Düster, A., Ruess, M., Garcia, E., Burgkart, R., Rank, E.: Nonstandard bone simulation: interactive numerical analysis by computational steering. Comput. Vis. Sci. **14**, 207–216 (2012)
51. Yang, Z., Ruess, M., Kollmannsberger, S., Düster, A., Rank, E.: An efficient integration technique for the voxel-based finite cell method. Int. J. Numer. Methods Eng. **91**, 457–471 (2012)

52. Zhu, T., Atluri, S.: A modified collocation method and a penalty formulation for enforcing the essential boundary conditions in the element free galerkin method. Comput. Mech. **21**, 211–222 (1998)
53. Zienkiewicz, O., Taylor, R.: The Finite Element Method – The Basis, vol. 1, 6th edn. Butterworth-Heinemann, Amsterdam/Boston (2005)

Immersed Boundary Methods for Fluid-Structure Interaction and Shape Optimization within an FEM-Based PDE Toolbox

Janos Benk, Hans-Joachim Bungartz, Miriam Mehl, and Michael Ulbrich

Abstract One of the main challenges in a classical mesh-based FEM-approach is the representation of complex geometries. This challenge is often tackled by a computationally costly mesh generation process, where the resulting mesh's facets represent the boundary. An alternative approach, that we employ here, is the immersed boundary (IB) approach. This uses instead a computationally cheaper structured adaptive Cartesian mesh and an explicit boundary representation, where the challenge mainly lies in the boundary condition (BC) imposition on the mesh cells intersected by the geometry's boundary. One IB method is Nitsche's method that we employ here for fluid-structure interaction (FSI) and shape optimization problems. The simulation of such complex physical systems modeled by PDEs requires a combination of sophisticated numerical methods. Implementing a FEM-based simulation software that computes a particular PDE's solution often requires the reusage of existing methods. In order to make our approach public and also to prove the modularity of it, we integrated our IB methods in an existing FEM-based PDE toolbox of the Trilinos project, called Sundance.

Keywords Fluid-structure interaction • Shape optimization • Immersed boundary methods

J. Benk (✉) · H.-J. Bungartz
Computer Science, Technische Universität München, Boltzmannstr. 3, 85748 Garching, Germany
e-mail: benk@in.tum.de; bungartz@in.tum.de

M. Mehl
Institute for Advanced Study, Technische Universität München, Lichtenbergstr. 2a, 85748 Garching, Germany
e-mail: mehl@in.tum.de

M. Ulbrich
Mathematics, Technische Universität München, Boltzmannstr. 3, 85748 Garching, Germany
e-mail: mulbrich@ma.tum.de

M. Bader et al. (eds.), *Advanced Computing*, Lecture Notes in Computational Science and Engineering 93, DOI 10.1007/978-3-642-38762-3_2,

1 Introduction

The idea behind the development of a finite element PDE toolbox suitable not only for simulation but also optimization algorithms is to provide a software allowing for the reuse of mesh functionality, data processing, matrix assembly, linear and non-linear solvers for various kinds of PDE based applications. To establish a new application code, the user only has to provide a suitable formulation of the continuous equations and of the finite element basis. Sundance [26, 27] has been developed with exactly that task. The user implements the particular application in a weak form formulation. This makes also the implementation of dual equations for optimization very convenient. In addition to the problem formulation, the type of finite element basis to be used has to be prescribed – of course in close connection with the computational grid and, therewith, the element type. Element quadrature, matrix assembly, and (non-)linear solvers are given by Sundance or Trilinos [20], respectively.

The focus of this paper is on the numerical simulation of fluid-structure interaction processes and related optimization problems. These problems require the flexible handling of complex and moving geometries including large geometrical or even topological changes. In particular for optimization, a high computational and memory efficiency is additionally required. Based on these requirements, we added a further grid type to the Sundance toolbox: structured adaptively refined Cartesian grids. Due to its strict structuredness, this grid type makes the storage of any explicit structure information such as relations between vertices, edges, faces, and elements or between elements and their neighbours obsolete and, thus, reduces the storage requirements to a minimum.

However, a major drawback of Cartesian grids is their inherently poor approximation accuracy for complex geometries. Thus, sophisticated methods to improve this accuracy are required. Here, immersed boundary methods such as Nitsche's method are a promising approach. In this paper, we present a Cartesian grid implementation in Sundance including an extension of Nitsche's method to flow simulations on moving geometries, in particular fluid-structure interaction and shape optimization problems.

In the following section, we show the underlying equations of our fluid-structure scenarios and recall some basic methods for PDE constraint optimization. Section 3 gives a an overview of immersed boundary methods and the related terminology. In Sect. 4, we shortly introduce the Sundance toolbox with its basic functionality, user interface, and implementational concept. We explain the Cartesian grid implementation in Sect. 5 and introduce Nitsche's method for fluid dynamics in moving geometries (Sect. 6). Section 7 presents details on the implementation in Sundance, whereas numerical examples for fluid-structure interactions using Nitsche's method and their results are shown in Sect. 8. For shape optimization, we present first simple scenarios with results in Sect. 9. Finally, we summarize results and give an outlook on future work in the conclusion (Sect. 10).

2 The Fluid-Structure Interaction Model and Optimization Methods

The Fluid-Structure Interaction Model. For the simulation of fluid-structure interactions on Cartesian grids, we use the incompressible laminar Navier-Stokes equations

$$\frac{\partial \mathbf{v}}{\partial t} = \nu \Delta \mathbf{v} - (\mathbf{v} \cdot \nabla)\mathbf{v} - \nabla p + \mathbf{f}_f \text{ (momentum equation)}, \tag{1}$$

$$\nabla \cdot \mathbf{v} = 0 \text{ (continuity equation)} \tag{2}$$

to model the fluid flow in connection with the structural dynamics equations

$$\rho_s \frac{\partial^2 \mathbf{u}}{\partial t^2} + \nabla \cdot \sigma_s + \mathbf{f}_s = 0, \text{ (structural dynamics)} \tag{3}$$

where $\mathbf{v}$ denotes the flow velocities, p the fluid pressure, $\mathbf{f}_f$ external forces acting on the fluid such as gravity forces, and ν the viscosity of the considered fluid. Further, $\mathbf{u}$ is the structure displacement vector, ρ_S the structure density, σ_s the stress tensor, and $\mathbf{f}_s$ denotes external traction forces exerted on the structure.

The interaction between fluid and structure is given by the continuity of velocities and the balance of forces

$$\mathbf{v} = \frac{\partial \mathbf{u}}{\partial t}, \tag{4}$$

$$\sigma_f \mathbf{n}_f = \sigma_s \mathbf{n}_s, \tag{5}$$

where σ_f and σ_s are the stress tensors of fluid and structure and $\mathbf{n}_f$ and $\mathbf{n}_s$ denote the normal vectors of the fluid and structure domain boundaries.

The structure is usually simulated in a Lagrangian framework, i.e., the grid deforms with the structure movement, such that the surface of the structure is always represented accurately. For the fluid domain, we, however, use an Eulerian fixed grid setting to allow also for large geometry or even topology changes.

Shape Optimization Methods. In our example applications, we consider designing the shape of a body B exposed to Navier-Stokes flow such that a given cost functional is minimized. We denote this functional with $J(y, \Omega)$ (e.g., the drag) under the constraints $E_\Omega(y) = 0$ (Navier-Stokes equations including boundary conditions), where $y = (\mathbf{v}, p)$ denotes the flow variables and Ω is the fluid domain surrounding the body to be optimized. Additional constraints such as constant volume of the body shape or smoothness of the body's boundary ensure the existence of an optimal design. A particular challenge of shape optimization is the fact that during the iterative process of optimization a sequence of different domains Ω_k is generated and the PDE has to be solved on each of these changing domains. This is closely related to boundary movement due to structure deformation in the fluid-structure interaction applications described above.

For gradient-based optimization, the derivative of $j(\Omega) := J(y(\Omega), \Omega)$ with respect to domain variations $\Omega \mapsto (\mathrm{Id} + V)(\Omega)$ is required, where V is a displacement field and Id the identity operator.

A classical approach, the method of mappings [6, 18, 32], works with a reference domain Ω_{ref} and bi-Lipschitz transformations $T : \mathbb{R}^n \to \mathbb{R}^n$ to represent the current domain via T as $\Omega = T(\Omega_{\mathrm{ref}})$. Using this transformation, the problem can be transported to the fixed domain Ω_{ref}. This results in a nonlinear PDE constrained optimal control problem with T serving as the control (design parameterization). If preferred, it can be arranged that the k-th optimization iteration performs its computations on the current domain Ω_k by viewing it as the reference domain. For conventional FEM, this requires moving the computational mesh as well as remeshing if the moved mesh deteriorates.

A closely related alternative to the method of mappings is provided by shape differential calculus [42,47]. It offers a rich machinery for computing the directional derivative $dj(\Omega)[V] := \frac{d}{dt} j((\mathrm{Id}+tV)(\Omega))\big|_{t=0}$, called shape derivative if it exists in the Gâteaux sense, in the displacement direction V. The Hadamard-Zolesio structure theorem [42, 47] states that under suitable conditions there exists a distribution g on the boundary of Ω such that $dj(\Omega)[V] = \langle g, V \cdot n\rangle_{\partial\Omega}$. The distribution g represents the sensitivity of $j(\Omega)$ with respect to normal boundary variations. It can be used to develop shape optimization algorithms that move the boundary of Ω to achieve a descent method.

The shape gradient g or a discrete version of it can be obtained by the adjoint approach [21]. In contrast to the sensitivity method, the computational complexity of the adjoint approach is independent of the dimension of the considered space of boundary displacement directions V. It requires to compute the adjoint state. If we denote by $\mathbf{u}$ the state and abbreviate the PDE as $E(\mathbf{u}, \Omega) = 0$, then the adjoint state $\mathbf{w}$ solves the adjoint PDE

$$(E_{\mathbf{u}}(\mathbf{u}, \Omega)\mathbf{v}, \mathbf{w}) = -(J_{\mathbf{u}}(\mathbf{u}, \Omega), \mathbf{v}) \quad \forall\, \mathbf{v} \in U_0, \tag{6}$$

where U_0 is the space of state variations and the subscript denotes the partial Fréchet derivative of the respective operator or function.

3 Immersed Boundary Methods: A Short Overview

The term immersed boundary methods summarizes a class of methods aiming at the accurate description of complex and moving geometries in a fixed Cartesian grid. It was developed already in the 1970s by Peskin [36] for the simulation of blood flow in the human heart. The basic idea is to embed a moving structure in a fluid flow simulated on a fixed (Eulerian) Cartesian grid. The influence of the structure on the flow is taken into account by a force function (which is zero away from the structure). A good overview on immersed boundary methods is given in [29].

In literature, there exists a variety of methods and a lot of names besides immersed boundaries that are related to the same topic: handling complex and moving geometries in fixed and mostly Cartesian grids. We shortly mention and classify the most important ones:

Marker-and-Cell. Early marker-and-cell methods, introduced by Harlow and Welch already in the 1960s [19] use cell-markes to defined fluid and non-fluid cells in a free surface flow. This approach obviously can easily be extended to complex flow geometries with obstacles [16] and to fluid-structure interaction [12]. However, the accuracy is restricted by the character of the grid, i.e., for Cartesian grids to $O(h)$ if h is the mesh width.

Volume-of-Fluid. To increase accuracy, volume-of-fluid methods have been introduced by Noh and Woodward in 1976 [34]. Instead of simply assigning cells the values zero for an empty cell and one for a cell completely filled with fluid, the volume-of-fluid method introduces a fractial function giving the fraction (in terms of volume) of a grid cell which is filled with fluid. This method has the advantage to allow a fulfillment of mass conservation. The accuracy of the actual geometry representation is only first order for the original method, but can be enhanced to second order (see, e.g., [37]).

Level-Set. To overcome the drawbacks of the volume-of-fluid method, another 12 years later, Osher and Sethian introduced the level-set method that tracks the fluid surface using a function that is zero at the domain boundary (and positive inside the flow domain) [35]. This allows even for topology changes in a natural way, conserves mass if applied correctly, and allows for high order of accuracy. The transient computation of the level-set function requires the solution of a Hamilton-Jacobi equation, which is neither trivial nor computationally cheap. The fractial function of the volume-of-fluid method can be interpreted as a volume integral over the level-set function.

Cut Cell. In contrast to the aforementioned approaches, cut cell methods [25] work with an explicit representation of the domain boundary by, e.g., polygons in 2D, triangulations in 3D, or higher-order surface representation with splines, for example. Classical cut cell methods use a finite volume discretization on the resulting geometry consisting of 'standard volumes' and 'cut cell volumes'. Accordingly, the roots of cut cell methods come from flow simulation. The accuracy depends on the accuracy of the boundary representation. A drawbacks of the cut cell method in particular in 3D is the large variety of possible types of cuts through a cell, and, therewith, the large number of different cut cell types. The advantage is the easy geometry representation with straightforward possibilities to interface with CAD (Computer Aided Design), e.g.

Fictitious Domain. Also fictitious domain methods [10] have been developed originally for fluid dynamics but are applied as well for structural dynamics [38]. The complex geometry is embedded in a simpler domain such as a square in 2D or a cube in 3D, and the original equation is replaced by an equation on this simpler domain by introducing jumping coefficients. These jumping coefficients can be interpreted as an extremely high or low stiffness outside the actual domain if the domain is surrounded by a rigid body or vacuum, e.g. The resulting system can be discretized either by finite differences, finite volumes, or finite elements. The drawback of the fictitious domain method in this simple form is that it allows only the modeling of rather simple boundary conditions such as homogeneous Dirichlet or Neumann boundary conditions. As soon as other boundary conditions such as those originating from a constant movement of a rigid body in a fluid have to be applied, additional efforts are required such as adding a Lagrange multiplier to enforce the boundary condition (see, e.g., [15]). The accuracy is limited by the accuracy of the discrete representation of the location of coefficient jumps as well as the influence of averaging effects when discretizing the underlying PDE at such a jump.

Penalty Methods. The penalty method proposed by Babuška in the late 1960s and early 1970s [1,2] for the Poisson equation with homogeneous boundary conditions allows to weakly enforce boundary conditions with a finite element basis that is not conforming with these boundary conditions. It enhances the energy functional to be minimized by the finite element solution by a boundary penalty term $h^{-s} \int_{\partial\Omega} v^s dS$. In case of a partial differential equation $L(u) = f$ with a differential operator L, this gives:

$$F(v) = \int_{\Omega} L(v)v - 2fv\,d\Omega + h^{-s} \int_{\partial\Omega} v^s dS. \tag{7}$$

This methods performs well in practice. One can be shown to be sensitive with respect to the choice of s and to deteriorate the order of convergence of the original discretization in theory.

Nitsche's Method. Nitsche developed a closely related method [33], that also enhances the energy functional in a finite element setting by a boundary penalty term, but adds further terms ensuring the symmetry of the resulting discrete equations and, therewith ensures full consistency. Nitsche's method has been applied to computational fluid dynamics in [4] and [3]. In the latter, the method is called *weakly enforced boundary conditions*.

Extended Finite Elements. In contrast to penalty and Nitsche's methods, extended finite element methods [30] enhance the finite element basis (enriched basis functions) at sharp *internal* boundaries, where discontinuities in material properties or cracks cause discontinuities in the numerical solution of a partial differential equation.

In this paper, we describe an extension of Nitsche's method for transient problems with moving boundaries (Sect. 6). This is new and requires some additional techniques to ensure the numerical stability of the time-stepping scheme.

4 The Sundance Toolbox in a Nutshell

The Sundance toolbox is a convenient software providing the user with a vast functionality for several pieces of the so-called simulation pipeline: problem formulation, i.e., mesh generation and weak form formulation of the required partial differential equations, discretization including assembling of a finite element matrix, computation of element matrices and stiffness matrix assembly, and, finally, solvers for the resulting systems of linear or nonlinear discrete equations. Figure 1 shows the steps of the simulation pipeline and outlines the contributions from users, Sundance components and external software components are linked to Sundance via well-defined interfaces.

A computational grid in Sundance is considered as a collection of mesh entities (elements, faces, edges, vertices) and their relations. A mesh compatible with the Sundance interface has to provide the following functionalities:

1. Return the unique identification number and dimensionality (0 for vertices, 1 for edges, 2 for faces, 3 for 3D elements) of a mesh entity,
2. Return the position, i.e., the spatial coordinates of a given vertex,
3. Return all lower-dimensional components of a mesh entity, e.g., all faces, edges, and vertices of a three- dimensional element,
4. Return the indices of all mesh elements that contain a given mesh entity together with its indices within the respective elements.

The grid can either be generated by an external mesh generator and read via a Sundance compatible parser from an input file or be generated internally within Sundance given the characteristic parameters of the mesh. Whereas the first approach is very well suited for unstructured grids and assures a high flexibility with respect to geometry and mesh generation with existing user-trusted tools, the latter is more suited for grids with a fixed structure such as our Cartesian grids. The mesh functionality described above can be either implemented by getting the required information from stored data (as for unstructured grids) or deriving information from the given structure of the grid.

For parallel simulations, the mesh interface is enhanced by functions tracing the assignment of mesh entities to processes:

1. Return the ID of the owner process of a mesh entity,
2. Transform the local ID of a mesh entity to the local ID on the respective process (and vice versa).

The domain partitioning itself can again be done via an interface to external established mesh partitioners or be implemented in Sundance itself.

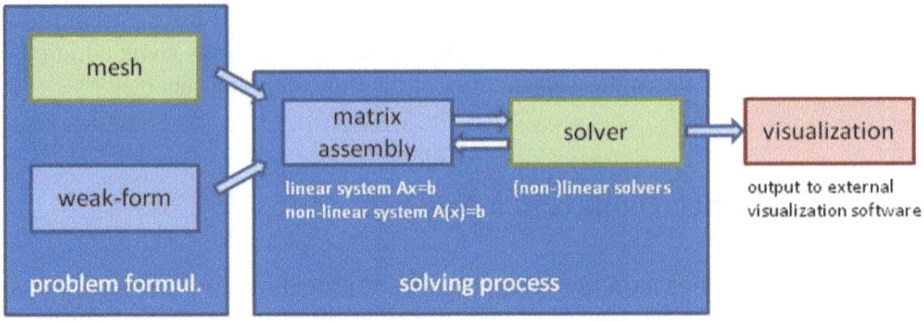

Fig. 1 Steps of the simulation pipeline in Sundance. Functionalities provided by Sundance are marked *light blue*, functionalities provided both by Sundance and by interfaces to external tools are *green*, and functionalities provided exclusively by interfaces to external software are marked *red*

As mentioned above, the equations describing the application are formulated by the user in a weak-form syntax. We give an example from [7] to demonstrate how easy a new application can be implemented in Sundance:

- We want to solve the two-dimensional Poisson equation, which is given by the weak form

$$\int_{\Omega} (\nabla \mathbf{u} \nabla \mathbf{v} - \mathbf{f}\mathbf{v})\, d\Omega = 0 \text{ for all } \mathbf{v} \text{ in } V.$$

- We start with reading a mesh that has been generated and written to a file by an external mesh generator:

```
MeshType meshType = new BasicSimplicialMeshType();
MeshSource meshReader =
  new TriangleMeshReader("meshInputFile.1", meshType);
Mesh meshExternal = meshReader.getMesh();
```

- Subsequently, we formulate the equations to be solved:

```
CellFilter Omega = new MaximalCellFilter();
CellFilter Boundary = new BoundaryCellFilter();

Expr unknBase = new Lagrange(1);
Expr testBase = new Lagrange(1);
Expr u = new UnknownFunction( unknBase , "u");
Expr v = new TestFunction( testBase , "v");
Expr dx = new Derivative(0);
Expr dy = new Derivative(1);
Expr grad = List(dx, dy);
QuadratureFamily quad = new GaussianQuadrature(2);
Expr weakForm =
  Integral( Omega , (grad*u)*(grad*v) - f*v, quad);
Expr bc = EssentialBC( Boundary , v*(u-1.0), quad);
```

- Finally, we setup the discrete problem and solve it with a suitable solver, which we choose from Trilinos [20]:

```
LinearProblem prob(mesh, weakForm, bc, v, u, vecType);

ParameterXMLFileReader reader("bicgstab.xml");
ParameterList solverParams = reader.getParameters();
LinearSolver<double> solver =
  LinearSolverBuilder::createSolver(solverParams);

Expr up = prob.solve(solver);
```

The example above shows a simple stationary linear example. For time-dependent applications, the time-stepping has to be defined by the user, in addition. Similarly, optimization problems require the formulation of the dual equations and the optimization algorithm by the user. Sundance does not implement an own visualization functionality but offers output functions writing the results in several standard formats suitable for different standard softwares such as VTK [41], ExodusII [40], and Matlab [28].

5 Cartesian Grids and Their Implementation in Sundance

Introduction of and General Remarks on Cartesian Grids. In contrast to unstructured grids, tree-structured adaptive Cartesian grids have the advantage of being highly efficient in terms of storage requirements. Whereas for an unstructured grid all elements, faces, edges, nodes, and, in particular, their relations have to be stored explicitly, our Cartesian grids are defined by the chosen tree-structure such that the only information that has to be stored is whether a grid cell is further refined or not. The grid on the left in Fig. 2 is for example given by the bit sequence

$$1\ 0\ 1\ 0000\ 0\ 0,$$

where the first '1' denotes the root, i.e., the whole square domain, the subsequent '0' stands for the non-refined lower left cell at refinement level one, the '1' for the refined lower right cell, followed by four '0's for its children and two '0's for the non-refined upper left and upper right level one cells. Thus, in this example, the bitsequence is ordered in a depth-first manner with Morton order of the cells at each level. It shows that an arbitrary tree-structured adaptive Cartesian grid can be stored with a memory requirement of only one bit per grid cell (over all levels) if three important parameters defining the type of grid and the traversal order are given:

1. The refinement factor (in our example two per coordinate direction),
2. The type of tree-traversal (depth-first or breadth-first),
3. The traversal order of cells at the same level (in our example Morton order).

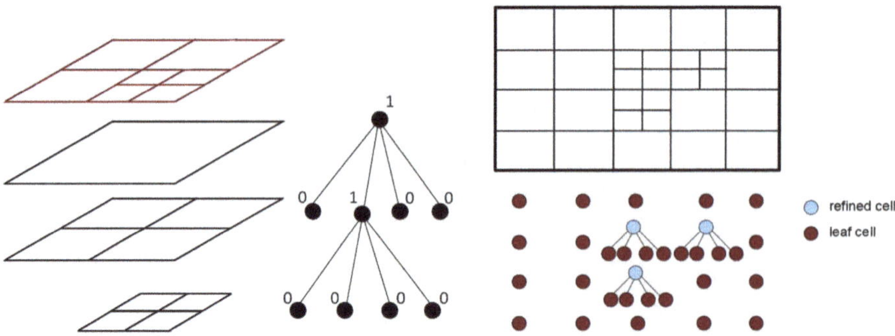

Fig. 2 Simple examples for tree-structured adaptive Cartesian grids represented either by a single cell-tree (*left*) or a forest of trees such as in p4est [13] and our parallel implementation of adaptive Cartesian grids in Sundance (Pictures taken from [7])

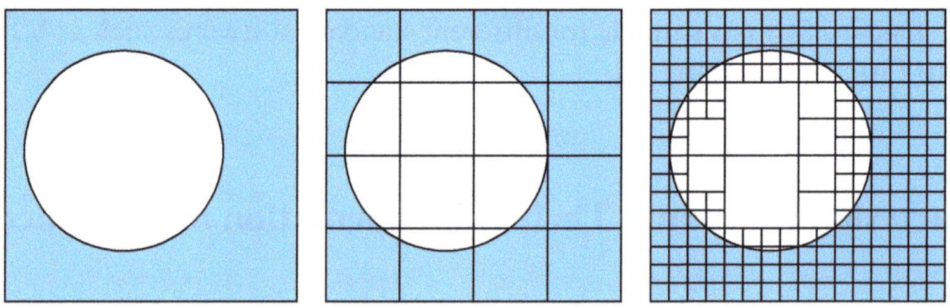

Fig. 3 Quadtree grid generation for a fluid domain around a spherical obstacle. Grid cells are further refined in this example if they are either cut by the geometry boundary or inside the fluid domain. This results in a regular refinement in the fluid domain. For visualization purposes, we show only every second refinement step. The whole grid generation corresponds to only one traversal of the final grid or its corresponding cell tree, respectively

Similar efficiency arguments hold for the generation of an adaptive Cartesian grid in an arbitrary domain: Starting from a square or cube containing the whole computational domain, the grid is locally further refined wherever a refinement is required for an accurate discretization of the partial differential equation to be solved or for a good approximation of the domain boundary. Figure 3 shows an adaptively refined grid representing a fluid domain around a spherical obstacle. This example also shows the straight-forward adaptivity process that can follow geometry information, i.e., intersection of cells by the domain boundary, adaptivity criteria, or a priori user information on domains of particular interest. If the grid is implemented using sophisticated data algorithms, adaptivity can even be done in a way that does not destroy data locality [45].

An additional advantage of adaptive Cartesian grids for application scenarios with moving geometries has already been mentioned in the introduction: a high flexibility for large geometry or even topology changes in an Eulerian, i.e., fixed grid setting. For these cases, we use immersed boundary methods to approximate

Table 1 Storage requirements of a three-dimensional Poisson solver using our implementation of a tree-structured Cartesian grid and an unstructured grid with a resolution of 531,441 ($81 \times 81 \times 81$) quad elements (results from [7])

Mesh	Mesh storage (MB)	Total memory (MB)
Cartesian mesh	783	1,600
Unstructured mesh	2,700	3,200

the geometry, the imposed boundary conditions and the differential operators in grid cells intersected by the domain boundary with suitable accuracy. Our immersed boundary approach is described in Sect. 6 in more detail. For the parallelization or domain partitioning of tree-structured adaptive Cartesian grids, space-filling curves have been shown to be a very cheap tool still providing quasi-minimal domain surfaces and, thus, communication costs [11, 17].

Mesh Generation and Storage in Sundance. A tree-structured Cartesian mesh can be stored in a highly efficient way as illustrated above. However, the Sundance mesh interface allows in principle for random access to mesh entities according to their IDs. A memory-saving tree-storage of the grid can not efficiently answer such arbitrary queries as each query would require a whole tree-traversal in the worst case. As a compromise, we linearize the tree and store all information of faces, edges, and vertices of all grid entities. As we can still use constant (up to scaling) element matrices and need only few interpolation matrices for elements with hanging nodes, we still save a considerable amount of memory if compared to unstructured grids. Table 1 compares the memory requirements of a Poisson solver with linear elements on tree-structured Cartesian and unstructured simplex meshes.

Table 1 shows that fitting a structured mesh into an arbitrary access context of a PDE framework usually deteriorates the theoretical memory saving potential. In a tree-traversal oriented implementation such as in Peano [45], the memory requirements are 50–100 times lower. However, grid generation is still done in a very simple and efficient way for our structured grids in a recursive refinement process, where the grid generator only needs to be given the information whether to locally refine further or not. This information can be derived from the geometry (refine only cells intersected by the domain boundary, e.g.), from error estimators, or from a priori given user input. The saving factor in terms of runtime strongly depends on the unstructured mesh generator and the scenario, but is obviously expected to be tremendously high for complex examples. Just remember that generating a Cartesian tree-structured grid corresponds to only a single grid traversal.

Introducing Rectangular Elements in Sundance. Sundance originally provided only simplex meshes. Thus, the first step towards adaptive Cartesian grids was the implementation of an additional, rectangular element including degrees of freedom located at the cell's vertices, edges, and faces. As finite element basis functions, we use the classical Lagrangian polynomial. The definition of a new basis only requires to implement the Sundance interface for basis function evaluation returning the basis function value for any given point on the reference element: In addition, the spatial derivatives are computed by automated differentiation.

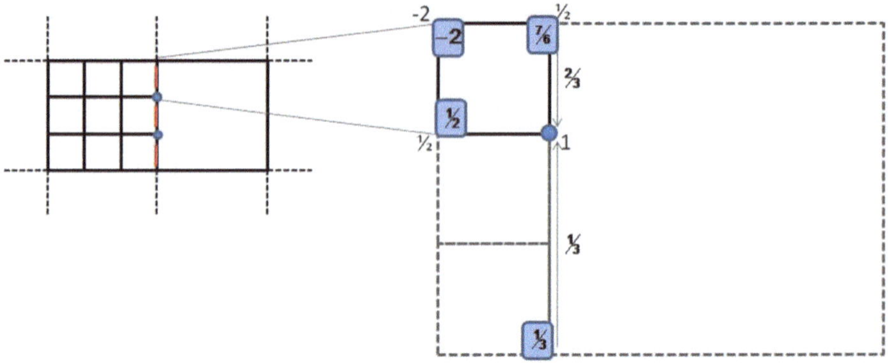

Fig. 4 Pre-fill transformation for a hanging node in a two-dimensional grid. *Left*: simple two-dimensional grid with hanging nodes (*blue*) and hanging faces (*red*); *right*: pre-fill transformation for element matrix lines associated to non-hanging element nodes interpolating the value at the hanging node from non-hanging father nodes. The resulting entries to be accumulated to the global system matrix are marked with *blue rectangles*

Hanging Nodes and Pre-Fill Transformation. In contrast to unstructured meshes, adaptive Cartesian meshes contain lower-dimensional mesh components (nodes, edges, faces), that are not a node, edge, or face for all neighbouring elements. Figure 4 shows a two-dimensional example with hanging nodes and edges. To ensure continuity of the approximate solution, hanging nodes, edges, or faces may not possess degrees of freedom and, accordingly, do not have associated entries in the vector of unknowns approximating the solution of our PDE. However, for efficiency reasons, we use the same element matrices (up to suitable scaling with the mesh width) in all cells. Thus, element matrices using a value at a hanging node, e.g., have to be modified as illustrated in Fig. 4 using an interpolation corresponding to the used finite element basis of the hanging node value from neighbouring non-hanging father nodes. The interpolation information is stored in a square matrix $M_{e,B}$ describing the interpolation of global (non-hanging) degrees of freedom to local (including hanging) values. In the example of Fig. 4, matrix $M_{e,B}$ for the upper left cell would be

$$M_{e,B} = \begin{pmatrix} 1 & 0 & 0 & 0 \\ 0 & \frac{1}{3} & 0 & \frac{2}{3} \\ 0 & 0 & 1 & 0 \\ 0 & 0 & 0 & 1 \end{pmatrix} \tag{8}$$

for bilinear basis functions and a lexicographic ordering of cell vertices: the first line indicates that the value of the local degree of freedom at the lower left vertex of the cell is the same as the value of the corresponding global degree of freedom (this node is not hanging). The same holds for the two upper vertices (line 3 and 4 of $M_{e,B}$). From line 2 of $M_{e,B}$, we see that we get the local value (at the hanging node in the lower right corner of the cell) as an averaged mean of the corresponding global value (lower right corner of the father cell) and the value at the upper right corner of the cell itself.

Independent from the actual grid, only a constant and rather small number of different interpolation matrices exists. Thus, we store them in a set of matrices and add only the index of the respective matrix $M_{e,B}$ (see Eq. (8)) to an element e in the mesh that contains hanging nodes, edges, or faces. Mathematically, the transformation of the element matrix A_e reads

$$A_e^{new} = M_{e,B_T}^T A_e M_{e,B_U}, \tag{9}$$

where B_T denotes the basis of the test space and B_U the basis of the ansatz space. This can be seen as an additional step of the matrix assembly process to be executed before accumulating element matrix contributions to the global system matrix. Therefore, we call this step the pre-fill transformation, which can be done very fast and efficiently due to the low number of cases and the purely local and small transformation matrices.[1] For the realization of the pre-fill transformation, the Sundance mesh interface sketched shortly in Sect. 4 has been extended by the following functionalities:

1. Return the information, if a face, edge, or vertex is hanging,
2. Return the required face, edge, or vertex of the father cell for interpolation,
3. Return the index of the cell within the parent cell.

The index of a cell within its father cell is required to determine the correct interpolation weights and indices of father cell entities.

Parallel Implementation of Cartesian Grids. The parallel implementation of a Cartesian grid has to provide two main ingredients: (1) a load balanced domain partitioning and (2) ghost layers for communication reduction. Once having these two ingredients, all other aspects such as the technical realization of data communication is taken care of by Sundance. For (1), we use the simple Z- or Morton-curve for grid partitioning. To further speed up the computations, we do the partitioning on a coarse grid level ignoring further refinement levels. This of course can lead to severe imbalances for grids with highly localized refinement and, in these cases, has to be substituted by a partitioning taking into account all grid levels. Figure 5 shows a very simple two-dimensional example for the partitioning of a grid into two subdomains.

Due to our pre-fill transformation algorithm, the second step, the determination of ghost cells, is a little more involved. We have to include also cells that do not have a face, edge, or vertex on the subdomain boundary for the reason illustrated in Fig. 5: grid elements at the subdomain boundary might have hanging facets and

[1] Note that we restrict our grids to 1-irregular tree-structured grids, which ensures that, if a face, edge, or vertex is hanging, the corresponding face, edge, or vertex of the father cell is not hanging.

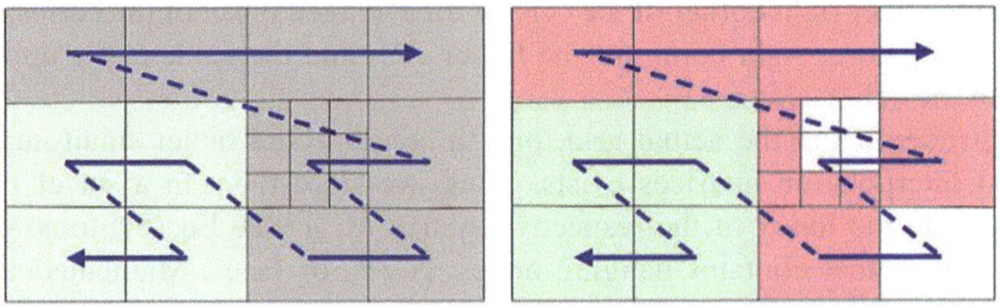

Fig. 5 *Left*: grid partitioning of a part of a very simple two-dimensional adaptive Cartesian grid into two subdomains using the Z-curve (Morton order) on a coarse grid level. *Right*: One of the two partitions (*light green*) with its required ghost cells (*pink*). Due to the pre-fill transformation, not only cells with a face on the subdomain boundary are required as ghost cells (Illustration taken from [7])

Table 2 Strong scaling results for a parallel three-dimensional solver of the Stokes-Brinkmann equation for porous media flow with varying porosity in a complex geometry with spherical low-porosity obstacles. The computations were performed on an adaptively refined Cartesian grid with 433, 126 cells and $Q_1 Q_1$ elements (see right subfigure for a cut through the geometry and the grid) on the MPP cluster of the Leibniz Supercomputing Center (LRZ) in Garching (Opteron 2.6 GHz AMD processors). As a system solver, we used the TSF-BiCGStab solver from the Trilinos [20] library

Number of proc.	1	2	4	8	16	32	64	128
Assembly (s)	682	356	183	98	52	28	18	10
Par. eff. (%)	100	96	93	87	82	76	59	53
Solver (s)	875	497	238	136	91	49	36	21
Par. eff. (%)	100	88	92	80	60	56	38	33

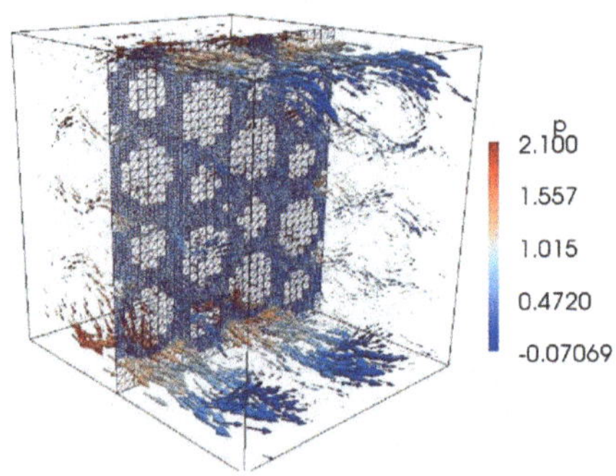

therefore contribute to the system matrix entries of neighbouring cells on a coarser level together with other child cells of the same father cell, which then might need information of a face of an additional coarser level neighbouring cell and so forth. However, the determination of ghost cells can still be done in a straightforward way by following such dependencies. In the current implementation, we still store the whole grid information in every process, which obviously causes a too high memory overhead and has to be improved (currently work in progress). Table 2 shows strong scaling results for a porous media flow in a complex domain. It shows the scalability of the matrix assembly process for our Cartesian grid and the solver runtime, where matrix assembly scales better than the solver (TSF-BiCGStab solver from the Trilinos [20] library), which proves the adequacy of our implementation.

After having the technical basis of Cartesian grids in the Sundance toolbox, we proceed with the description of the Nitsche method, which is the decisive step making the Cartesian grid implementation useful for accurate and efficient simulations.

6 Nitsche's Method for Fluid Dynamics in Moving Geometries

The general idea of Nitsche's method is the following: Let Γ_d be the Dirichlet boundary of $\Omega \in \mathbb{R}^n$ with boundary values g. This usually implies the state space $U_g = \{u \in U; u|_{\Gamma_d} = g\}$ and the space of test functions is $W_0 = \{w \in W; w|_d = 0\}$ with appropriate function spaces U and W. In this case, Dirichlet conditions are built into the function space U_g. The weak form is obtained by testing the strong equation with the test function and performing integration by parts where appropriate. The resulting boundary integrals usually vanish on Γ_d since $w|_{\Gamma_d} = 0$.

The Nitsche method works differently: The spaces U and W are used as state space and test function space and are oblivious of the boundary values g. Now, doing integration by parts, the boundary integral over Γ_d does no longer drop out. These integrals over Γ_d are symmetrized, i.e., if we have a summand

$$\int_{\Gamma_d} l\left(\partial_{\mathbf{n}} u, w\right) dS(x)$$

with a bilinear functional l, we replace it by

$$\int_{\Gamma_d} l\left(\partial_{\mathbf{n}} u, w\right) dS(x) + \int_{\Gamma_d} l\left(\partial_{\mathbf{n}} w, u\right) dS(x) - \int_{\Gamma_d} l\left(\partial_{\mathbf{n}} w, g\right) dS(x). \tag{10}$$

Note that for the exact solution u of our partial differential equation with Dirichlet boundary values g, the added terms cancel.[2] Furthermore, we add regularizations, also called penalty terms

$$\frac{\gamma}{h} \int_{\Gamma_d} uwdS(x) - \frac{\gamma}{h} \int_{\Gamma_d} gwdS(x) \tag{11}$$

[2]The symmetrization is required to show that the error of the finite element solution minimizes a quadratic energy functional, which is used to prove the consistency order of the method (compare [33]).

and possibly

$$\frac{\gamma_2}{h}\int_{\Gamma_d}(u\cdot\mathbf{n})(w\cdot\mathbf{n})dS(x)-\frac{\gamma_2}{h}\int_{\Gamma_d}(g\cdot\mathbf{n})(w\cdot\mathbf{n})dS(x) \tag{12}$$

to coercify the problem and to enforce the Dirichlet conditions.

Nitsche's Method for the Incompressible Navier-Stokes Equations. For the formulation of Nitsche's method for the incompressible Navier-Stokes equations, we use the notation of Becker [4]. Let $\Omega\subset\mathbb{R}^2$ (or $\Omega\subset\mathbb{R}^3$) and $\mathbf{u}=(\mathbf{v},p)\in H^1(\Omega)^2\times L_0^2(\Omega)$ (or $H^1(\Omega)^3\times L_0^2(\Omega)$) be the state space (velocity and pressure). Further, let $\Gamma=\partial\Omega$, $\mathbf{g}\in H^{1/2}(\Gamma)$, and $\mathbf{f}\in L^2(\Omega)^2$ (or $L^2(\Omega)^3$). We consider the stationary Navier-Stokes equations in a first step:

$$-\nu\Delta\mathbf{v}+(\mathbf{v}\cdot\nabla)\mathbf{v}+\nabla p=\mathbf{f}\text{ in }\Omega, \tag{13}$$

$$\nabla\cdot\mathbf{v}=0\text{ in }\Omega, \tag{14}$$

$$\mathbf{v}=\mathbf{g}\text{ at }\Gamma. \tag{15}$$

In the following, we restrict to 2D for simplicity, although all methods are applicable also in 3D. For deriving a weak formulation, we test the PDE with $\Phi=(\Psi,\xi)\in H^1(\Omega)^2\times L_0^2(\Omega)$ and, as usual, integrate two terms by parts:

$$\begin{aligned}(-\nu\Delta\mathbf{v},\Psi)&=\nu\int_\Omega-[\nabla\cdot(\nabla\mathbf{v})]\cdot\Psi dx=\nu\int_\Omega\nabla\mathbf{v}:\nabla\Psi dx-\nu\int_\Gamma\partial_{\mathbf{n}}\mathbf{v}\cdot\Psi dS(x)\\&=\nu\left(\nabla\mathbf{v},\nabla\Psi\right)-\nu\left\langle\partial_{\mathbf{n}}\mathbf{v},\Psi\right\rangle, \qquad (16)\\(\nabla p,\Psi)&=-\int_\Omega p(\nabla\cdot\Psi)dx+\int_\Gamma p\mathbf{n}\cdot\Psi dS(x)=-\left(p,\nabla\cdot\Psi\right)+\langle p\mathbf{n},\Psi\rangle, \qquad (17)\end{aligned}$$

where $(\cdot,\cdot)$ denotes the scalar product introduced by the domain integral, $\langle\cdot,\cdot\rangle$ the scalar product introduced by the boundary integral. Summing up (16), (17), and the weak form of the convective term in (13), we thus get the usual distributed terms

$$a(\mathbf{u},\Phi)=\nu(\nabla\mathbf{v},\nabla\Psi)+((\mathbf{v}\cdot\nabla)\mathbf{v},\Psi)-(p,\nabla\cdot\Psi)-(\nabla\cdot\mathbf{v},\xi), \tag{18}$$

the right hand side term $(\mathbf{f},\Psi)$, and the new left hand side boundary terms

$$c(\mathbf{u},\Psi)=-\nu\langle\partial_{\mathbf{n}}\mathbf{v},\Psi\rangle+\langle p\mathbf{n},\Psi\rangle. \tag{19}$$

For symmetrization, we add $c(\Phi,\mathbf{v})$ on the left and compensate this by adding $c(\Phi,\mathbf{g})$ on the right (analogue to (10)). For stabilization and for enforcing the Dirichlet conditions (compare (11) and (12)), we add

$$\nu\frac{\gamma_1}{h}\langle\mathbf{v},\Psi\rangle+\frac{\gamma_2}{h}\langle\mathbf{v}\cdot\mathbf{n},\Psi\cdot\mathbf{n}\rangle$$

on the left and compensate this by adding

$$\nu \frac{\gamma_1}{h} \langle \mathbf{g}, \Psi \rangle + \frac{\gamma_2}{h} \langle \mathbf{g} \cdot \mathbf{n}, \Psi \cdot \mathbf{n} \rangle$$

on the right. Becker [4] uses a further stabilizing term $-\langle (\mathbf{vn})^- \mathbf{v}, \Psi \rangle$, where $(\mathbf{vn})^-$ equals $\mathbf{vn}$ if $\mathbf{vn} < 0$ and zero else. Taking all together, we have

$$\begin{aligned} & a(\mathbf{u}, \Phi) + c(\mathbf{u}, \Psi) + c(\Phi, \mathbf{v}) + \nu \tfrac{\gamma_1}{h} \langle \mathbf{v}, \Psi \rangle + \tfrac{\gamma_2}{h} \langle \mathbf{v} \cdot \mathbf{n}, \Psi \cdot \mathbf{n} \rangle - \langle (\mathbf{vn})^- \mathbf{v}, \Psi \rangle \\ & = (\mathbf{f}, \Psi) + c(\Phi, \mathbf{g}) + \nu \tfrac{\gamma_1}{h} \langle \mathbf{g}, \Psi \rangle + \tfrac{\gamma_2}{h} \langle \mathbf{g} \cdot \mathbf{n}, \Psi \cdot \mathbf{n} \rangle - \langle (\mathbf{gn})^- \mathbf{g}, \Psi \rangle \text{ for all } \Phi. \end{aligned} \tag{20}$$

If we consider the instationary Navier-Stokes equations

$$\frac{\partial \mathbf{v}}{\partial t} = \nu \Delta \mathbf{v} - (\mathbf{v} \cdot \nabla)\mathbf{v} - \nabla p + \mathbf{f} \text{ in } \Omega, \tag{21}$$

$$\nabla \cdot \mathbf{v} = 0 \text{ in } \Omega, \tag{22}$$

$$\mathbf{v} = \mathbf{g} \text{ at } \Gamma, \tag{23}$$

we have to add the time-derivative to the weak-form formulation. Doing this in the classical way, i.e., not using a space-time finite element approach, yields

$$\begin{aligned} & \left(\tfrac{\partial \mathbf{v}}{\partial t}, \Psi \right) + a(\mathbf{u}, \Phi) + c(\mathbf{u}, \Psi) + c(\Phi, \mathbf{v}) + \nu \tfrac{\gamma_1}{h} \langle \mathbf{v}, \Psi \rangle + \tfrac{\gamma_2}{h} \langle \mathbf{v} \cdot \mathbf{n}, \Psi \cdot \mathbf{n} \rangle - \langle (\mathbf{vn})^- \mathbf{v}, \Psi \rangle \\ & = (\mathbf{f}, \Psi) + c(\Phi, \mathbf{g}) + \nu \tfrac{\gamma_1}{h} \langle \mathbf{g}, \Psi \rangle + \tfrac{\gamma_2}{h} \langle \mathbf{g} \cdot \mathbf{n}, \Psi \cdot \mathbf{n} \rangle - \langle (\mathbf{gn})^- \mathbf{g}, \Psi \rangle \text{ for all } \Phi. \end{aligned} \tag{24}$$

For time-discretization, we can use any finite difference scheme such as Euler or Runge-Kutta.

Issues with Moving Geometries. Whereas the generalization of Nitsche's method to transient problems is straight-forward, the treatment of problems with moving or changing geometries is more involved in our fixed grid setting. To demonstrate the principle problem, we consider an explicit Euler time-stepping scheme applied to the Nitsche formulation (24):

$$\begin{aligned} \left(\mathbf{v}^{(k+1)}, \Psi\right) = {} & \left(\mathbf{v}^{(k)}, \Psi\right) + dt \cdot \left[-\tilde{a}(\mathbf{u}^{(k)}, \mathbf{u}^{(k+1)}, \Phi) - c(\mathbf{u}^{(k)}, \Psi) - c(\Phi, \mathbf{v}^{(k)}) \right. \\ & -\nu \frac{\gamma_1}{h} \langle \mathbf{v}^{(k)}, \Psi \rangle - \frac{\gamma_2}{h} \langle \mathbf{v}^{(k)} \cdot \mathbf{n}, \Psi \cdot \mathbf{n} \rangle + \langle (\mathbf{v}^{(k)} \mathbf{n})^- \mathbf{v}^{(k)}, \Psi \rangle \\ & + (\mathbf{f}^{(k)}, \Psi) + c(\Phi, \mathbf{g}^{(k)}) + \nu \frac{\gamma_1}{h} \langle \mathbf{g}^{(k)}, \Psi \rangle + \frac{\gamma_2}{h} \langle \mathbf{g}^{(k)} \cdot \mathbf{n}, \Psi \cdot \mathbf{n} \rangle \\ & \left. - \langle (\mathbf{g}^{(k)} \mathbf{n})^- \mathbf{g}^{(k)}, \Psi \rangle \right] \text{ for all } \Phi, \end{aligned} \tag{25}$$

where

$$\tilde{a}(\mathbf{u}^{(k)}, \mathbf{u}^{(k+1)}, \Phi) = \nu(\nabla \mathbf{v}^{(k)}, \nabla \Psi) + ((\mathbf{v}^{(k)} \cdot \nabla)\mathbf{v}^{(k)}, \Psi) - (p, \nabla \cdot \Psi) + (\nabla \mathbf{v}^{(k+1)}, \xi).$$

Evaluating (25) for the basis functions $\Phi_i = (\Psi_i, 0)$ and $\Phi_{N+i} = (0, \xi_i), i = 1, \ldots, N$ leads to the system of equations for time step $t^{(k)} \rightarrow t^{(k+1)}$. We get the explicit Euler with Chorin's projection method for the pressure by replacing $\mathbf{v}^{(k+1)}$ in the last N equations (discrete continuity equation) by the formula given in the first N equations (discrete momentum equations). As we compute a solution at time step $t^{(k+1)}$, we have to integrate expressions involving $\mathbf{v}^{(k)}$ over $\Omega^{(k+1)}$. However, $\mathbf{v}^{(k)}$ does not provide physically meaningful values in $\Omega^{(k+1)} \setminus \Omega^{(k)}$. To achieve a consistent formulation, one possibility would be to apply space-time finite element formulations [5, 43]. This, however, would require four-dimensional cut-cell quadrature functionality for spatially three-dimensional problems. Thus, we solve the moving geometry problem with a simpler approach in a first step: We restrict ourselves to fully implicit time-stepping schemes and, thus, minimize the impact of values of $\mathbf{v}^{(k)}$ in $\Omega^{(k+1)} \setminus \Omega^{(k)}$. In addition, we prevent $\mathbf{v}^{(k)}$ from taking extreme values outside $\Omega^{(k)}$ by solving a Poisson equation with a small weighting factor in the fictitious domain (compare [7]). The results presented in Sects. 8 and 9 show the usability of this simple approach for a suite of fluid-structure interaction benchmarks from [22, 23] and for shape optimization.

7 Implementation in Sundance

To implement Nitsche's method for flow equations on complex and moving domains, we first need some technical ingredients, i.e., volume integrals over cut cells integrating only over the part of a grid element that is inside the fluid domain and boundary integrals over the domain boundary. Both require an explicit representation of the domain boundary. As we work with second-order discretizations, we can use simple polygons (in 2D)[3] or surface triangulations (in 3D). Figure 6 illustrates the representation of domains and cut cell volumes. In 2D, we allow a cell to be intersected by an arbitrary polygon. In this case, we can always find a suitable decomposition of the remaining fluid part of a cell into simple trapezoidals and triangles. In 3D, the analogue process with arbitrary surface triangulations intersecting a cubic grid element would lead to a too large amount of different cases and, thus, is not feasible. Therefore, we restrict to cases, where all cell edges are intersected at most once by the surface triangulation. Connecting these intersection points by straight lines leads to a new, approximate, but still second-order accurate surface triangulation and only a small number of cases as displayed in Fig. 6. In both cases, 2D and 3D, we end up with a decomposition of the fluid part of the cut cell into simple elements $E_i, i \in I_E$ of given type that can be integrated up to machine precision with suitable quadrature rules. The same holds for the surface integrals.

[3]For 2D geometries, there is also an implementation in Sundance that can deal with analytical geometry representations (such as a circle) that are then automatically approximated by a suitable polygon internally [7].

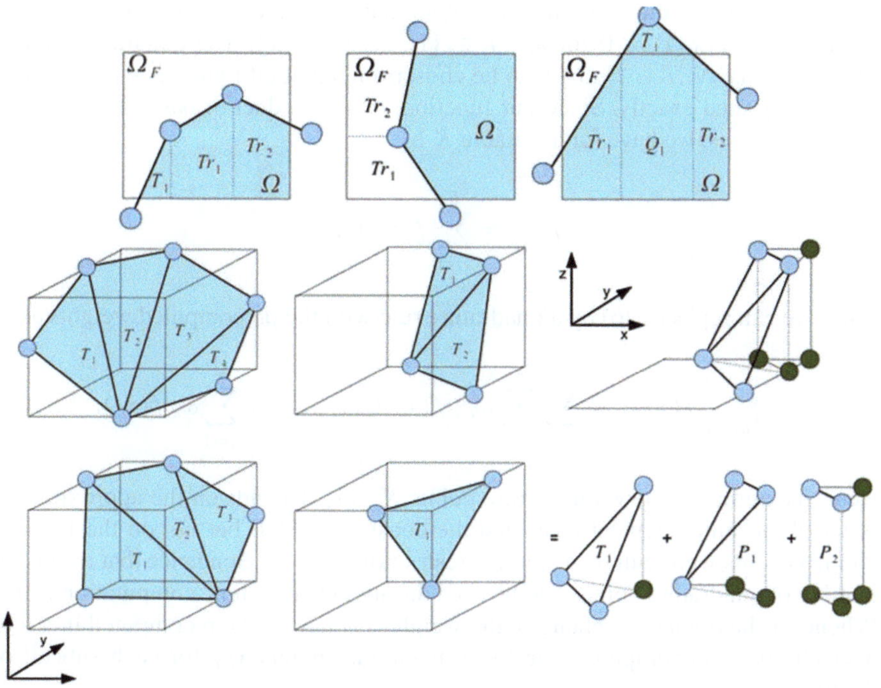

Fig. 6 Illustration of the decomposition of the geometry boundary and the fluid parts of a cut cell in two-dimensional and three-dimensional cases. This method allows for the computation of the finite element integrals in cut cells and on boundaries described by polygons in 2D and triangulations in 3D with only a small set of quadrature rules for simple geometric elements (Illustrations taken from [7])

For the volume integrals over cut cells, we thus have

$$\int_{E\cap\Omega} f(x)dx = \sum_{i\in I_E} \int_{E_i} f(x)dx = \sum_{i\in I_E} \sum_{j=1}^{N_i} \omega_{i,j} f(x_{i,j})dx, \tag{26}$$

where $E_i, i \in I_E$, are the simple subcells into which we decompose of the fluid part of the cell and $x_{i,j}, j = 1, \dots, N_i$, are the quadrature points used in subcell i. With this formula, we can compute all cut cell integrals up to machine precision. However, this method can lead to a large number of quadrature points $x_{i,j}$ in particular in 3D and for higher order basis functions. Therefore, we reduce the number of quadrature points by using precomputed weights

$$\omega_k = \sum_{i\in I_E} \sum_{j=1}^{N_i} \omega_{i,j} l_k(x_{i,j}), k = 1, \dots, K \tag{27}$$

where l_k denotes the Lagrangian polynomial associated to the point x_k, i.e., $l_k(x_k) = 1, l_k(x_m) = 0$ for all $m \in \{1, \dots, K\} \setminus \{k\}$. The quadrature rule in (27) and, thus, $N_i, i \in I_E$, have to be chosen such that all Langrange polynomials l_k are integrated exactly. Since our function f is a product of basis functions and their derivatives, we have for a suitable K by

$$f(x) = \sum_{k=1}^{K} f(x_k) l_k(x).$$

Thus, we can replace (26) by a quadrature rule with the precomputed weights ω_k:

$$\int_{E \cap \Omega} f(x)dx = \sum_{i \in I_E} \sum_{j=1}^{N_i} \omega_{i,j} f(x_k) l_k(x_{i,j}) dx = \sum_{k=1}^{K} \omega_k f(x_k). \tag{28}$$

Here, the number of new quadrature points K only depends on the approximation order of the finite element basis and the considered PDE but *not* on the function $f(x)$. This fits well with the technical restriction given by Sundance that a constant number of quadrature points is to be used throughout the whole computational grid. Whenever the domain Ω changes, the weights ω_k have to be recomputed in a loop over all cut cells computing the local quadrature points $x_{i,j}$ for each subcell on the fly.

Line or surface boundary integrals are computed analogously. However, we do not use precomputed weights here as (1) some boundary integrals involve normal vectors that are different for each subpart of the boundary inside a given cell, (2) boundary integrals are cheaper such that we can afford storing all quadrature points, (3) three-dimensional cases require the usage of data points in the vicinity but not directly on the boundary due to the approximation of the original triangulation within each cut cell.

Technically, the implementation of Nitsche's method in Sundance requires new user interfaces for

- Description of the geometry by a polygon/triangulation
- Cell filters selecting intersected cells,
- Integrals over cut cells, and
- Boundary integrals.

To realize moving boundaries in Sundance, several data have to be updated after changing of the geometry:

- The positions of the surface polygon or triangulation,
- Internal information such as IDs of cells cut by components of the surface object,
- Precomputed quadrature weights for the finite element integration during matrix assembly,
- Entries of the system matrix/matrices.

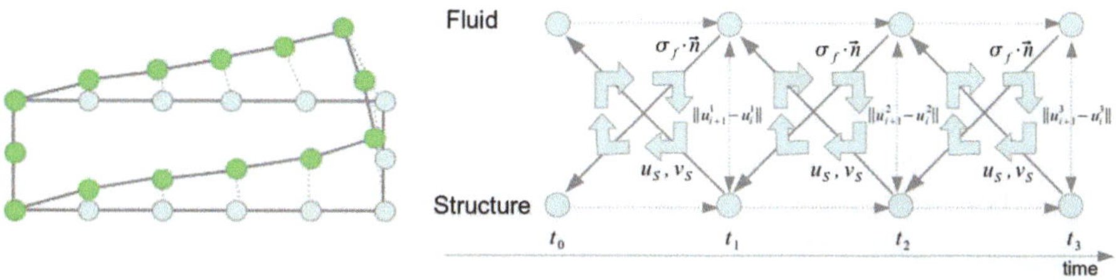

Fig. 7 *Left*: Eulerian (fluid) and Lagrangian (structure) view of a simple two-dimensional moving structure. The *green circles* mark the nodes of the polygonal geometry description in the fluid solver, whereas the *light grey circles* are the boundary vertices of the Lagrangian reference domain on the structure side. *Right*: Schematic view of the staggered coupling approach for partitioned fluid-structure interaction simulations. After executing a time step of the flow solver, forces at (29) are transfered to the structure solver, which subsequently executes its time step and produces displacements and velocities at the wet surface as an output, which serve as a boundary condition at (29) for the next fluid step and so forth. This process is repeated until the residual of the fix-point equation for the displacements falls below a given threshold (Illustrations taken from [7])

To move surface nodes, we use the local surface velocity (given by the boundary conditions) at the respective position. The update of all internal geometry information is hidden from the user in the function .update(). The function .reAssembleProblem() recomputes quadrature weights and assembles the new system matrix/matrices.

8 Fluid-Structure Simulations Using Nitsche's Method

The suitability and accuracy of our method using Cartesian grids in combination with Nitsche's method for simple model problems and pure flow simulation in static and moving domains has been shown in [7, 8]. In this paper, we focus on results obtained for a multiphysics application class, i.e., fluid-structure interactions.

Our Partitioned Approach for Coupled Fluid-Structure Simulation in Sundance. We use the incompressible Navier-Stokes equations (1) and (2) and combine them with the structural equation (3) and the coupling conditions (4) and (5) to a full fluid-structure simulation environment.

As we use an Eulerian fixed grid approach as described in the previous sections for the fluid part, but the common Lagrangian approach with a classical enforcement of boundary conditions for the structure part, we have to couple the interface displacements in addition:

$$\Gamma_F = \{x + \mathbf{u}_s | x \in \Gamma_L\}, \tag{29}$$

where Γ_F is the internal moving geometry description (polygon or surface triangulation) of the fluid solver and Γ_L the Lagrangian structure surface description (constant throughout the simulation). See Fig. 7 for an illustration. Technically, this

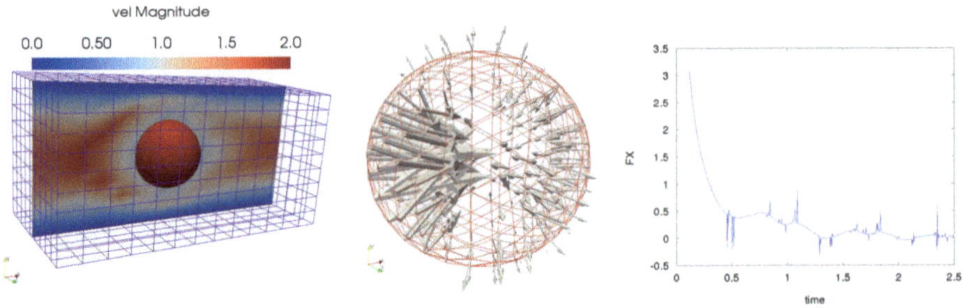

Fig. 8 Simulation of a rigid body motion of a sphere in a fluid flow using a very coarse Cartesian mesh. *Left*: absolute velocities in a plane of the fluid domain; *middle*: force vectors acting on the sphere; *right*: total forces in flow direction acting on the sphere plotted over time. The forces decay to zero as the sphere accelerates (Illustrations taken from [7])

is realized in Sundance using twin polygons or twin triangulations. The nodes of these two polygons or triangulations are connected via a bijective mapping from the node set of the first polygon/triangulation to the node set of the second polygon/triangulation as illustrated in Fig. 7 by the dashed lines. Whenever a value of an unknown function is changed on a node of one polygon/triangulation, it is automatically also changed on the respective node of the twin polygon/triangulation.

As a time discretization, we use implicit Euler for the fluid and structure equations. To perform a time step of the whole coupled problem, the partitioned approach requires an iterative execution of fluid and structure solver steps involving an exchange of boundary values. We use the widespread staggered coupling approach as shown in Fig. 7. For incompressible fluids, this coupling in general leads to unstable time stepping if used in an explicit manner (only one fluid and structure step per time step). Thus, we have to iteratively repeat the fluid and structure steps until the residual of the associated fixed point equation for the displacements at the wet surface falls below a certain threshold. To achieve convergence of the fluid-structure iterations, we use Aitken underrelaxation when transferring displacements from the structure solver to the flow solver. Aitken underrelaxation determines an adaptive scalar underrelaxation factor in each iteration [24]. More sophisticated coupling methods such as interface quasi-Newton schemes [14] or multilevel [9,46] approaches yield an even faster convergence. However, for our purpose of showing the potential of Nitsche's method for the simulation of fluid-structure interaction on structured Cartesian grids, Aitken underrelaxation suffices.

In the following, we present some of our numerical results for fluid-structure interaction scenarios in 2D and 3D. Further results can be found in [7].

Rigid Body Motion in 3D. Our first scenario is the three-dimensional rigid body motion of a spherical object in a flow channel as shown in Fig. 8. In a channel of dimensionless size $8 \times 4 \times 4$, a sphere with dimensionless radius one and a dimensionless mass of one moves according to Newton's law of motion and the sum of forces exerted on it by the fluid flow. The fluid has the dimensionless density

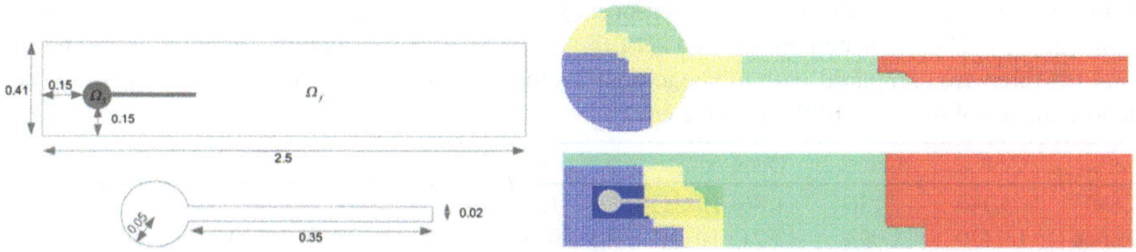

Fig. 9 Two-dimensional fluid-structure interaction benchmark scenario from [22, 23]. The scenario shows a fixed and rigid spherical object with an attached super-elastic compressible flap structure embedded in a channel flow. *Left*: Scenario geometry with a slight offset of the structure from the channel axis. *Right*: domain decomposition of the structure and the fluid part for a parallel simulation with four processors (Illustrations taken from [7])

one and the inflow speed is set to 2.25. We keep the sphere fixed until dimensionless time $T = 0.2$. After that initial period, the sphere is released and accelerated by the surrounding fluid flow. The fluid flow is simulated on a very coarse regular Cartesian mesh with only $13 \times 9 \times 9$ elements, whereas the sphere's surface is approximated by 500 triangles with 252 nodes. Due to the relatively large structure density and the rather small radius of the sphere, this example allowed for explicit coupling, i.e., we performed only one fluid and structure step in each time step. Figure 8 shows the good solution quality in spite of the very coarse fluid mesh resolution. The sphere exhibits a smooth movement in flow direction as expected for a correct solution. The forces in x-direction show some unphysical peaks due to the time integration issues for moving geometries mentioned above. Depending on the state of the simulation, these peaks can even result in negative forces slowing down the motion of the sphere in x-direction. However, each of these peaks decays very fast and, thus, does not influence the overall solution significantly. The height of the peaks obviously depends on the time step size and the spatial resolution.

Two-Dimensional Benchmarks. To verify our implementation by more reliable data than heuristic observations as in the 3D example above, we simulated 2D benchmark scenarios for fluid-structure interaction proposed in [22, 23]. The scenario consists of a channel flow with an embedded spherical obstacle and an attached flexible bar structure. The spherical obstacle itself remains fixed, but the bar bends in a steady state (FSI1) or oscillating (FSI3) manner depending on the Reynolds number of the channel flow. Due to a slight offset in vertical direction, the structure is exposed to both horizontal and vertical forces. Figure 9 shows the geometric setup of the benchmark. The structure is a super-elastic compressible material with $\rho_s = 10^3 \frac{\text{kg}}{\text{m}^3}$ and Young's modulus $E = 1.4 \cdot 10^6 \frac{\text{kg}}{\text{m}^3}$. We discretize the Navier-Stokes equations using Q_2Q_1 and the structure using Q_1 elements. For this scenario, an implicit coupling with several fluid-structure iterations is mandatory to achieve stable results. We use $\epsilon = 10^{-6}$ as a stopping criterion for our staggered fluid-structure iterations with a constant underrelaxation factor $\omega = 0.3$.

The FSI1 scenario is a stationary case with a steady state flow and a slight constant bending of the elastic bar structure in y-direction. The inflow velocity is

Table 3 Simulation results for two different mesh resolutions and the benchmark scenario FSI1 from [22, 23]. The displacement of the tip of the bar in x- and y-direction, the total drag, and the total lift force are compared with the benchmark reference values in [23]. Numbers in parentheses denote the absolute value of the relative error

Cells_{fluid}	Cells_{struct}	Lines_{poly}	Displ. x	Displ. y	Drag	Lift
2,610	3,380	136	$1.86 \cdot 10^{-5}$(18 %)	$1.230 \cdot 10^{-3}$(50 %)	14.078(1.5 %)	0.8202(7.4 %)
36,666	13,370	1,093	$2.21 \cdot 10^{-5}$(2.6 %)	$0.703 \cdot 10^{-3}$(14 %)	14.198(0.7 %)	0.8101(6.1 %)
54,104	13,370	1,093	$2.18 \cdot 10^{-5}$(3.5 %)	$0.846 \cdot 10^{-3}$(5.2 %)	14.224(0.5 %)	0.7930(3.8 %)
Reference values			$2.27 \cdot 10^{-5}$	$0.821 \cdot 10^{-3}$	14.295	0.7638

set to $v_f = 0.2 \frac{\text{m}}{\text{s}}$. Convergence within a time step was achieved after 20 iterations on average. Table 3 shows the computed values for the displacements of the tip of the bar in x- and y-direction as well as the total drag and lift forces exerted on the structure by the fluid. The total forces have been calculated by curve integrals on the polygon. The table shows already a good agreement with the benchmark values for the coarsest mesh resolution and a convergence towards the correct values with increasing resolution. The simulations for this scenario have been performed in parallel using four processors for each, fluid and structure solver, and two processors for the surface polygon. The associated domain decomposition of the fluid and structure domain is shown in Fig. 9. As a solver, we used the SuperLU-DIST linear solver within an outer non-linear NOX solver.[4] This combination resulted in an acceptable but not yet optimal parallel efficiency of 54 % [7]. Note that the simple load balancing strategy, a quasi-serial phase for the surface polygon, and further optimization potential in the storage of grid partitiones currently limits the scalability, a fact that shall be improved in further development steps.

The FSI3 is a transient case with a higher inflow velocity of $2 \frac{\text{m}}{\text{s}}$ that leads to an oscillatory movement of the flexible bar and also requires a strong, i.e., implicit coupling of fluid and structure. We used Aitken underrelaxation resulting in 11 iterations per time step on average. Figure 10 and Table 4 show the resulting movement of the bar and the respective tip displacements, total drag, and total lift with their mean value, the amplitude of their oscillation, and their frequency. The comparison with the benchmark reference values from [23] shows a good agreement. However, for more reliable convergence statements, additional mesh resolutions have to be examined in addition in future work.

Three-Dimensional Bending Tower. As a three-dimensional example for the coupled simulation of fluid-structure interactions with a moving flexible structure, we consider a channel flow with a bending tower (see Fig. 11). The fluid has a again density of $10^3 \frac{\text{kg}}{\text{m}^3}$ and a parabolic inflow profile with maximal velocity $0.45 \frac{\text{m}}{\text{s}}$. The bending tower is modeled by a super-elastic material with Young's modulus $E = 0.4 \cdot 10^6 \frac{\text{kg}}{\text{ms}^2}$. The tower has a size of $0.05 \times 0.25 \times 0.1$ m. We use an adaptive Cartesian grid for the flow solver with adaptive refinement in a rectangular domain

[4]trilinos.sandia.gov/packages/nox/

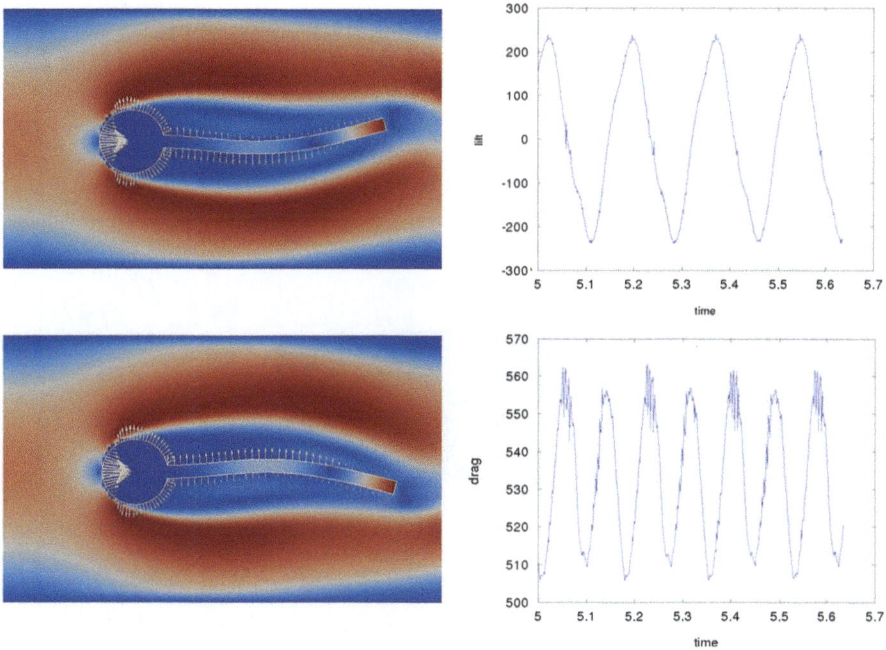

Fig. 10 Simulation results for the transient FSI3 benchmark from [22, 23]. *Left*: two snapshots of the solution with colours representing the absolute fluid velocities and the structural displacement. *Arrows* show the forces acting on the structure; *middle*: time variant total drag; *right*: time variant total lift (Plots taken from [7])

Table 4 Simulation results for two different mesh resolutions and the benchmark scenario FSI3 from [22, 23]. For all measured values, i.e., displacement of the tip of the bar in x- and y-direction, total drag, and total lift force, offset, amplitude, and frequency (in $[\cdot]$) are measured and compared with the benchmark reference values in [23]

$Cells_{fluid}$	$Cells_{struct}$	$Lines_{poly}$	Displ. x	Displ. y	Drag	Lift
2,507	864	136	−0.02866	−0.00111	524.5	56.50
			±0.02857[11.1]	±0.03301[5.5]	±23.5[11.1]	±214.50[5.5]
5,133	864	136	−0.00288	0.00166	532	−1.5
			±0.00282[11.4]	±0.03452[5.7]	±25[11.4]	±229.50[5.7]
Reference values			−0.00269	0.00148	457.3	2.2
			±0.00253[10.9]	±0.03438[5.3]	±22.66[10.9]	±149.78[5.3]

around the tower. The resolution of the coarse background mesh is $27 \times 7 \times 7$ cells. We discretized the Navier-Stokes equations with stabilized $Q_1 Q_1$ and the structure equation with Q_1 elements. The stabilized $Q_1 Q_1$ are known to work well in three-dimensional fluid benchmarks, e.g., in [39]. The coupling iterations are performed with an underrelaxation factor $\omega = 0.3$ up to an error tolerance $\epsilon = 10^{-6}$. Figure 11 shows the resulting bending of the tower, which can not be verified by benchmark values but gives reasonable results [7].

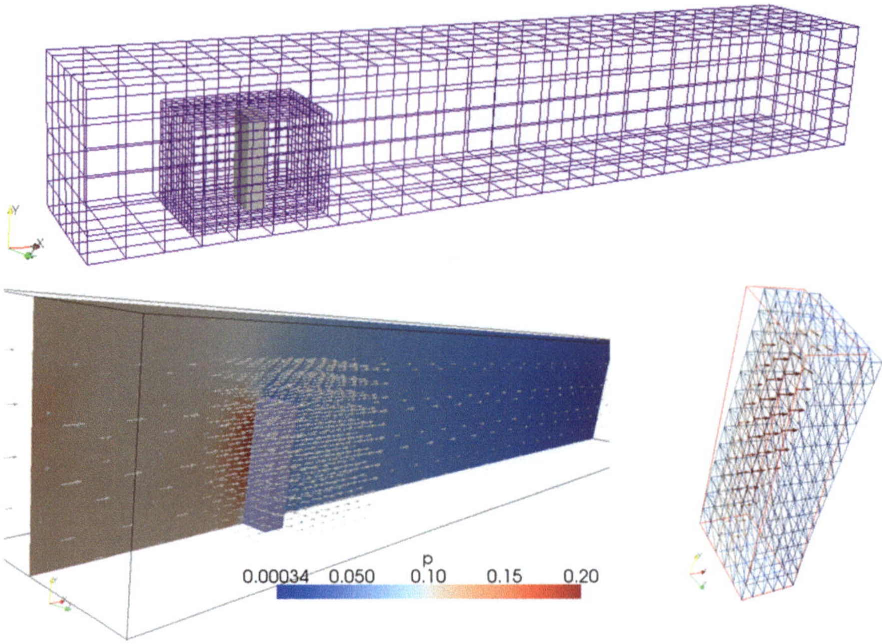

Fig. 11 Three-dimensional simulation of a bending tower in a channel flow. The *upper picture* shows the tower and the surrounding computational grid with an adaptive refinement in a block around the tower. *Left*: flow field and fluid pressure; *right*: bending of the tower (Illustrations taken from [7])

9 Shape Optimization with the Immersed Boundary Approach

We now demonstrate the application of our immersed boundary approach to shape optimization problems. To this end, we consider the Stokes equations

$$-\nu\Delta\mathbf{v} + \nabla p = 0, \quad \operatorname{div}\mathbf{v} = 0, \quad \mathbf{v}|_{\partial\Omega\setminus\Gamma} = d, \quad \mathbf{v}|_{\Gamma} = 0. \tag{30}$$

Here, $\Omega = \Omega_C \setminus B$, where Ω_C is a rectangular channel and B is a body with boundary Γ. The goal is to find a design for B such that the cost functional

$$J(\mathbf{v}, \Omega) = \nu \int_{\Omega} \|\nabla\mathbf{v}\|^2 \, dx, \tag{31}$$

which is related to the drag of the body, is minimized. Here, B shall have a prescribed volume and (except possibly at the front and rear tip points) a sufficiently smooth boundary. The latter condition is required to ensure existence of optimal shapes.

Denoting by $(\mathbf{v}(\Omega), p(\Omega))$ the solution of the Stokes equations on the domain Ω, the shape derivative of $j(\Omega) := J(\mathbf{v}(\Omega), \Omega)$ can be shown to satisfy

$$dj(\Omega)[V] = -\nu \int_{\Gamma} \|\partial_n \mathbf{v}(\Omega)\|^2 \, V \cdot n \, dS(x), \tag{32}$$

see [31, pp. 29/30]. A similar result holds for the more general case of the Navier-Stokes equations, but it then involves the adjoint state [31, pp. 31/32]. The Stokes equations in combination with the cost function (31) allow to express all occurrences of the adjoint state in terms of $\mathbf{v}$, thus making the adjoint state superfluous.

The shape of the body B is described by a closed polygon with nodes (x_0, y_0), ..., (x_{2m-1}, y_{2m-1}). Here, the next point (x_{2m}, y_{2m}) would correspond to (x_0, y_0). We only consider shapes that are symmetric with respect to a horizontal center line (parallel to the x-axis):

$$y_0 = y_m, \quad x_i = x_{2m-i}, \quad y_0 - y_i = y_{2m-i} - y_0 \quad (1 \leq i < m).$$

We do not work directly with this representation, but use polar coordinates with respect to $(a, b) = ((x_0 + x_m)/2, y_0)$. For $i = 0, \ldots, m$, the i-th point is allowed to move along the radial ray $(a, b) + \mathbb{R}(x_i - a, y_i - b)$. The transformed point is parameterized by the radial scaling r_i: $(x_i(r_i), y_i(r_i)) = (a, b) + r_i(x_i - a, y_i - b)$. Due to our symmetry assumptions, specifying $r = (r_0, \ldots, r_m)^T$ uniquely characterizes the shape of the polygonal boundary of the deformed body.

In the function space setting, the existence of optimal shapes requires coercivity of the cost function in a sufficiently strong shape space. In the discrete setting, we enforce some amount of smoothness by adding an H^1-like regularization for r to the cost function (31). Also, we work with an H^1-like inner product by switching to the parameterization $z \in \mathbb{R}^{m+1}$ with $r_i = \sum_{j=0}^{i} z_j$. We then have $z_0 = r_0$ and $z_i = r_i - r_{i-1}$, $i > 0$. In our computations, we work with the body $B(z)$ parameterized by z. We then can use an l_2-norm for the regularization:

$$\gamma m \sum_{i=1}^{m} z_i^2. \tag{33}$$

The factor m stems from the fact that if we view the r_i as evaluations of a function $r(\tau)$ on an equidistant grid $\tau_0, \ldots, \tau_m$ with $\tau_i = i\delta$, $\delta = 1/m$, then $m z_i = (r_i - r_{i-1})/\delta$ approximates $r'(\tau_i - \delta/2)$. Hence,

$$\int_0^1 r'(\tau)^2 \, d\tau \approx \delta \sum_{i=1}^{m} r'(\tau_i - \delta/2)^2 \approx m \sum_{i=1}^{m} z_i^2. \tag{34}$$

For evaluating the discrete version $j_h(z)$ of the cost function (31) and its gradient $\nabla j_h(z)$ at a point z in parameter space, we solve the Stokes equation, discretized by Nitsche's method, on $\Omega(z) = \Omega_C \setminus B(z)$ to obtain the discrete velocity field $\mathbf{v}_h$.

Then

$$J(\mathbf{v}_h, \Omega(z)) = \nu \int_{\Omega(z)} \|\nabla \mathbf{v}_h\|^2 \, dx \tag{35}$$

is computed and the regularization (33) is added. For evaluating the gradient $\nabla j_h(z)$ at a point z we use that the map $z \mapsto (x_i(z), y_i(z)) - (a, b)$ is linear. Thus, the direction of variation $e_i = (\delta_{ik})_{0 \leq k \leq m}$ of z is linearly mapped to the direction of variation $(x_p(e_i), y_p(e_i)) - (a, b)$ of the p-th polygon node $(x_p(z), y_p(z))$.

Therefore, for the z-variation direction e_i, the value of the corresponding displacement field direction $V = V^p$ at the polygon's pth node is $(x_p(e_i), y_p(e_i)) - (a, b)$. To compute $\nabla_{z_i} j_h(z)$ we thus evaluate

$$-\nu \int_{\Gamma(z)} \|\partial_n \mathbf{v}_h\|^2 \, V^p \cdot n \, dS(x). \tag{36}$$

The required methods, in particular the conversion of nodal values along the polygon into expressions that can appear in line integrands over the polygon curve have all been implemented by us for general use in Sundance.

We stress that the formula (36) for $\nabla_{z_i} j_h(z)$ is not fully exact since (32) requires the exact solution $\mathbf{v}$ of the PDE, whereas $\mathbf{v}_h$ is only a numerical approximation that, in addition, depends not only on the shape of $\Omega(z)$ but also on the underlying FEM mesh.

In the numerical test presented here we require also symmetry with respect to the vertical center line passing through (a, b). To this end, we choose $m = 2l$ and only prescribe the lower left quarter of the profile, which is described by $z_0, \ldots, z_l$, where we require $x_l = a$. Thus, (x_0, y_0) lies on the horizontal center line and (x_l, y_l) lies on the vertical center line. The rest of the profile is obtained by symmetry and the center of mass is then automatically fixed to (a, b).

We impose a constant volume constraint $\text{vol}(\Omega(z)) = c_{\text{vol}}$, with c_{vol} denoting the volume of the initial shape. The volume and its derivative can be obtained conveniently by, e.g., the Leibniz sector formula. We also pose inequality constraints on the maximum radial elongation/contraction: $1/2 \leq r_i(z) \leq 2$ for all i. The bounds are chosen such that the constraints do not become active, but they can help avoiding too strong initial design changes during optimization.

We choose $\Omega_C = (0, 2.5) \times (0, 1)$, $(a, b) = (0.75, 0.5)$, the tip point of the initial shape is at $(0.67, 0.5)$ and the initial volume is 0.0163672. We impose Dirichlet conditions $v = (0.2, 0)$ on the boundary of Ω_C and $v = 0$ on $\partial B(z)$. The boundary conditions on $\partial B(z)$ are enforced by Nitsche's method. Figure 12 shows the initial shape on the left. The boundary of the body is described by $2\,\text{m} = 4\,\text{l} = 80$ points, i.e., we have $m = 40$, $l = 20$. We use a Taylor-Hood discretization (biquadratic for the velocity, bilinear for the pressure) on a Cartesian mesh with hanging nodes. The grid consists of a 145×101 mesh, where the 23×41 subgrid centered around $(a, b) = (0.75, 0.5)$ is refined by cellwise trisection in both space dimensions to a obtain a 69×123 grid on this subregion. The regularization parameter is $\gamma = 0.006$.

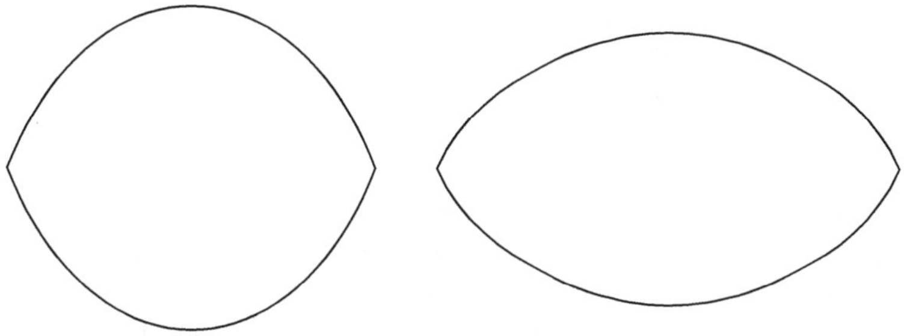

Fig. 12 Two-dimensional shape-optimization example for a flow around a rigid body. *Left*: initial shape; *right*: optimal shape

We implemented a general purpose C++ interface that couples our Sundance simulation, which solves the Stokes equation and evaluates j_h, ∇j_h, etc., to the interior point optimization software IPOPT [44]. We use the limited memory BFGS approximation of the Lagrange function's Hessian matrix that is built into IPOPT.

As mentioned, due to the fact that the shape derivative formula is only exact in the continuous setting, the computed derivatives are not exact derivatives of the discretized cost function. Thus, we can achieve only a moderate (but sufficient) accuracy in satisfying the stopping criteria of the NLP solver IPOPT.

Starting with the initial shape shown on the left in Fig. 12, we obtain a solution (with the achievable accuracy due to inexact discrete shape gradient) after 10 IPOPT iterations. The objective function is reduced by 6.9 %. The optimal shape is shown on the right in Fig. 12. Note that the result depends on the chosen regularization.

10 Conclusion

We proposed an implementation of adaptive Cartesian grids in combination with Nitsche's method for an accurate enforcement of boundary conditions on complex and moving domains. Adaptive Cartesian meshes have severe advantages for an efficient implementation of solvers for partial differential equations and PDE constraint optimization problems. These advantages reach from minimal memory requirements over efficient data access algorithms, easy load balanced domain decomposition using space-filling curves to a very efficient mesh generation and adaptive refinement process. In this paper, we implemented Cartesian grids in the Sundance toolbox designed for the fast, weak-form formulation based realization of PDE solvers and PDE constraint optimization algorithms. Sundance minimizes the user-effort for setting up a new application. In addition, it offers numerous solvers, e.g., by an interface to the Trilinos solver library. Thus, it is an ideal platform for the development and testing of finite element based methods for a higher-order representation of boundary conditions for Cartesian grids that inherently are not able to represent arbitrary complex and moving geometries with more than first order accuracy without further efforts. We used an old method known introduced by Nitsche in the 1970s that allows for a consistent weak enforcement of Dirichlet boundary conditions on boundaries cutting the Cartesian grid cells in an arbitrary way. We enhanced this method to flow simulations in moving geometries based on the incompressible Navier-Stokes equation and showed their accuracy even for cases of fluid-structure interactions, where the accuracy at the boundary between fluid and structure is particularly crucial, and for a shape-optimization example.

Fitting Cartesian grids to the software requirements of Sundance destroys some of their advantages. However, our implementation allows for a fast development and enhancement of methods such as the Nitsche method and, thus, lays the basis for a high-performance computing implementation. Future work is the investigation of methods limiting the condition number of system matrices resulting from Nitsche's method for very small cut cell fluid parts, the development of improved methods for moving geometries avoiding peaks in pressure or forces due to the extrapolation of previous time step solutions to the current time step's domain, and the examination of additional time-step restrictions induced by Nitsche's method.

References

1. Babuška, I.: Numerical Solution of Boundary Value Problems by the Perturbed Variational Principle. Technical Note BN-626, Institute for Fluid Dynamics and Applied Mathematics, University of Maryland, Oct 1969
2. Babuška, I.: The finite element method with penalty. Math. Comput. **27**, 122–128 (1973)
3. Bazilevs, Y., Michler, C., Calo, V.M., Hughes, T.J.R.: Isogeometric variational multiscale modeling of wall-bounded turbulent flows with weakly enforced boundary conditions on unstretched meshes. Comput. Methods Appl. Mech. Eng. **199**, 780–790 (2010)
4. Becker, R.: Mesh adaption for dirichlet flow control via nitsche's method. Commun. Numer. Methods Eng. **18**, 669–680 (2002)
5. Behr, M.: Simplex space-time meshes in finite element simulations. Int. J. Numer. Meth. Fluids **57**, 1421–1434 (2008)
6. Bello, J.A., Fernández-Cara, E., Lemoine, J., Simon, J.: The differentiability of the drag with respect to the variations of a Lipschitz domain in a Navier-Stokes flow. SIAM J. Control Optim. **35**(2), 626–640 (1997)
7. Benk, J.: Immersed boundary methods within a PDE toolbox on distributed memory systems. Ph.D. thesis, Technische Universität München (2012)
8. Benk, J., Mehl, M., Ulbrich, M.: Sundance PDE solvers on Cartesian fixed grids in complex and variable geometries. In: Proceedings of the ECCOMAS Thematic Conference CFD & Optimization, Antlya, 23–25 May 2011
9. Bijl, H., van Zuijlen, A.H., Bosscher, S.: Two level algorithms for partitioned fluid-structure interaction computations. In: Wesseling, P., Oñate, E., Périaux, J. (eds.) ECCOMAS CFD 2006, European Conference on Computational Fluid Dynamics. TU Delft, The Netherlands (2006)
10. Bugrov, A.N., Smagulov, S.: Fictitious domain method for Navier-Stokes equations. In: Mathematical Model of Fluid Flow, Novosibirsk, pp. 79–90 (1978)
11. Bungartz, H.J., Mehl, M., Weinzierl, T.: A parallel adaptive Cartesian PDE solver using space-filling curves. In: Nagel, E., Walter, V., Lehner, W. (eds.) Euro-Par 2006, Parallel Processing, 12th International Euro-Par Conference. Lecture Notes in Computer Science, vol. 4128, pp. 1064–1074. Springer, Berlin/Heidelberg (2006)

12. Bungartz, H.J., Gatzhammer, B., Mehl, M., Neckel, T.: Partitioned simulation of fluid-structure interaction on Cartesian grids. In: Bungartz, H.J., Mehl, M., Schäfer, M. (eds.) Fluid-Structure Interaction – Modelling, Simulation, Optimisation, Part II. LNCSE, vol. 73, pp. 255–284. Springer, Berlin/Heidelberg (2010)
13. Bursteddea, C., Wilcox, L., Ghattas, O.: p4est: scalable algorithms for parallel adaptive mesh refinement on forests of octrees. SIAM J. Sci. Comput. **33**(3), 1103–1133 (2011)
14. Degroote, J., Bathe, K., Vierendeels, J.: Performance of a new partitioned procedure versus a monolithic procedure in fluid-structure interaction. Comput. Struct. **87**, 793–801 (2009)
15. Glowinski, R., Pan, T., Hesla, T., Joseph, D., Périaux, J.: A fictitious domain approach to the direct numerical simulation of incompressible viscous flow past moving rigid bodies: application to particulate flow. J. Comput. Phys. **169**, 363–426 (2001)
16. Griebel, M., Dornseifer, T., Neunhoeffer, T.: Numerical Simulation in Fluid Dynamics, a Practical Introduction. SIAM, Philadelphia (1998)
17. Griebel, M., Zumbusch, G.: Hash based adaptive parallel multilevel methods with space-filling curves. In: Rollnik, H., Wolf, D. (eds.) NIC Symposium 2001, NIC Series, ISBN 3-00-009055-X, vol. 9, pp. 479–492, Forschungszentrum, Jülich (2002)
18. Guillaume, P., Masmoudi, M.: Computation of high order derivatives in optimal shape design. Numer. Math. **67**(2), 231–250 (1994)
19. Harlow, F.H., Welch, J.E.: Numerical calculation of time-dependent viscous incompressible flow of fluid with a free surface. Phys. Fluids **8**(12), 2182–2189 (1965)
20. Heroux, M., Bartlett, R., Howle, V., Hoekstra, R., Hu, J., Kolda, T., Lehoucq, R., Long, K., Pawlowski, R., Phipps, E., Salinger, A., Thornquist, H., Tuminaro, R., Willenbring, J., Williams, A.: An overview of trilinos. Technical Report SAND2003-2927, Sandia National Laboratories (2003). http://trilinos.sandia.gov/TrilinosOverview.pdf
21. Hinze, M., Pinnau, R., Ulbrich, M., Ulbrich, S.: Optimization with PDE constraints. In: Mathematical Modelling: Theory and Applications, vol. 23. Springer, New York (2009)
22. Hron, J., Turek, S.: Proposal for numerical benchmarking of fluid-structure interaction between elastic object and laminar incompressible flow. In: Bungartz, H.J., Schäfer, M. (eds.) Fluid-Structure Interaction. Lecture Notes in Computational Science and Engineering, vol. 53, pp. 371–385. Springer, New York (2006)
23. Hron, J., Turek, S.: Numerical benchmarking of fluid-structure interaction between elastic objects and laminar incompressible flow. In: Bungartz, H.J., Mehl, M., Schäfer, M. (eds.) Fluid-Structure Interaction – Modelling, Simulation, Optimisation, Part II. LNCSE, vol. 73. Springer, Berlin/Heidelberg (2010)
24. Kuettler, U., Wall, W.: Fixed-point fluid-structure interaction solvers with dynamic relaxation. Comput. Mech. **43**(1), 61–72 (2008)
25. LeVeque, R.: Cartesian grid methods for flow in irregular regions. In: Morton, K., Baines, M. (eds.) Numerical Methods in Fluid Dynamics, III, pp. 375–382. Clarendon Press, Oxford (1988)
26. Long, K.: Sundance 2.0 tutorial (2004). http://prod.sandia.gov/techlib/access-control.cgi/2004/044793.pdf
27. Long, K.R., Kirby, R.C., van Bloemen Waanders, B.: Unified embedded parallel finite element computations via software-based Frechet differentiation. SIAM J. Sci. Comput. **32**(6), 3323–3351 (2010)
28. MathWorks: MATLAB, The language of technical computing. http://www.mathworks.com/products/matlab/index.html. Last visted: Sept 2012
29. Mittal, R., Iaccarino, G.: Immersed boundary methods. Annu. Rev. Fluid Mech. **37**, 239–261 (2005)
30. Moës, N., Dolbow, J., Belytschko, T.: A finite element method for crack growth without remeshing. Int. J. Numer. Methods Eng. **46**, 131–150 (1999)
31. Mohammadi, B., Pironneau, O.: Applied Shape Optimization for Fluids. Oxford University Press, Oxford (2001)
32. Murat, F., Simon, J.: Etudes de problèmes d'optimal design. Lect. Notes Comput. Sci. **41**, 54–62 (1976)

33. Nitsche, J.: Über ein Variationsprinzip zur Lösung von Dirichlet-Problemen bei Verwendung von Teilräumen, die keinen Randbedingungen unterworfen sind. Abhandlungen aus dem Mathematischen Seminar der Universität Hamburg **36**, 9–15 (1971). http://dx.doi.org/10.1007/BF02995904. 10.1007/BF02995904
34. Noh, W.F., Woodward, P.: SLIC (Simple Line Interface Calculation). In: van de Vooren, A.I., Zandbergen, P.J. (eds.) Proceedings of 5th International Conference on Numerical Methods in Fluid Dynamics, Twente University. Lecture Notes in Physics, vol. 59, pp. 330–340 (1976)
35. Osher, S., Sethian, J.A.: Fronts propagating with curvature-dependent speed: algorithms based on Hamilton-Jacobi formulations. J. Comput. Phys. pp. 12–49 (1988)
36. Peskin, C.S.: Numerical analysis of blood flow in the heart. J. Comput. Phys. **25**(3), 220–252 (1977)
37. Pilliod, J., Puckett, E.: Second-order accurate volume-of-fluid algorithms for tracking material interfaces. J. Comput. Phys. **199**(2), 465–502 (2004)
38. Rank, E., Kollmannsberger, S., Sorger, C., Düster, A.: Shell finite cell method: a high order fictitious domain approach for thin-walled structures. Comput. Methods Appl. Mech. Eng. **200**(45/46), 3200–3209 (2011). doi:10.1016/j.cma.2011.06.005. http://www.sciencedirect.com/science/article/pii/S0045782511002234
39. Schäfer, M., Turek, S.: Benchmark computations of laminar flow around a cylinder. In: Flow Simulation with High-Performance Computers 2. Notes on Numerical Fluid Mechanics, vol. 52, pp. 547–566. Vieweg, Los Angeles (1996)
40. Schoof, L.A., Yarberry, V.R.: Exodus II: a finite element data model. Sandia report SAND92-2137, Sandia National Laboratories (1994)
41. Schroeder, W., Martin, K., Lorensen, W.: The design and implementation of an object-oriented toolkit for 3D graphics and visualization. In: Proceedings of the 7th Conference on Visualization, VIS'96, pp. 93ff. IEEE Computer Society Press, Washington, DC (1996)
42. Sokolowski, J., Zolésio, J.P.: Introduction to Shape Optimization: Shape Sensitivity Analysis. Springer, Berlin (1992)
43. Tezduyar, T.E., Behr, M., Mittal, S., Johnson, A.: Computation of unsteady incompressible flows with the stabilized finite element methods: space-time formulations, iterative strategies and massively parallel implementations. In: vol. 246, pp. 7–24. Asme Press, Vessel. Pip. Div. Publ. PVP (1992). http://www.tafsm.org/PUB_PRE/cALL/c21-PVPAMD92.pdf
44. Wächter, A., Biegler, L.T.: On the implementation of an interior-point filter line-search algorithm for large-scale nonlinear programming. Math. Program. **106**(1), 25–57 (2006)
45. Weinzierl, T., Mehl, M.: Peano – a traversal and storage scheme for octree-like adaptive cartesian multiscale grids. SIAM J. Sci. Comput. **33**(5), 2732–2760 (2011)
46. Yigit, S., Heck, M., Sternel, D.C., Schäfer, M.: Efficiency of fluid-structure interaction simulations with adaptive underrelaxation and multigrid acceleration. Int. J. Multiphysics **1**, 85–99 (2007)
47. Zolésio, J.P., Delfour, M.: Shapes and Geometries: Analysis, Differential Calculus and Optimization. SIAM, Philadelphia (2001)

Numerical Simulation of Transport in Porous Media: Some Problems from Micro to Macro Scale

Quanji Cai, Sheema Kooshapur, Michael Manhart, Ralf-Peter Mundani, Ernst Rank, Andreas Springer, and Boris Vexler

Abstract This paper deals with simulation of flow and transport in porous media such as transport of groundwater contaminants. We first discuss how macro scale equations are derived and which terms have to be closed by models. The transport of tracers is strongly influenced by pore scale velocity structure and large scale inhomogeneities in the permeability field. The velocity structure on the pore scale is investigated by direct numerical simulations of the 3D velocity field in a random sphere pack. The velocity probability density functions are strongly skewed, including some negative velocities. The large probability for very small velocities might be the reason for non-Fickian dispersion in the initial phase of contaminant transport. We present a method to determine large scale distributions of the permeability field from point-wise velocity measurements. The adjoint-based optimisation algorithm delivers fully satisfying agreement between input and estimated permeability fields. Finally numerical methods for convection dominated tracer transports are investigated from a theoretical point of view. It is shown that high order Finite Element Methods can reduce or even eliminate non-physical oscillations in the solution without introducing additional numerical diffusivity.

Keywords Porous media • Pore scale • High order FEM • Parameter identification

Q. Cai (✉) · R.-P. Mundani · E. Rank
Lehrstuhl für Computation in Engineering, Department of Civil Engineering and Surveying, Technische Universität München, Arcisstr, 21, 80333 München, Germany
e-mail: cai@bv.tum.de; mundani@tum.de; ernst.rank@tum.de

S. Kooshapur · M. Manhart
Fachgebiet Hydromechanik, Technische Universität München, 80333 München, Germany
e-mail: michael.manhart@tum.de; s.kooshapur@bv.tum.de

A. Springer · B. Vexler
Lehrstuhl für Opimale Steuerung, Fakultät für Mathematik, Technische Universität München, 85748 Garching b. München, Germany
e-mail: springer@ma.tum.de; vexler@ma.tum.de

M. Bader et al. (eds.), *Advanced Computing*, Lecture Notes in Computational Science and Engineering 93, DOI 10.1007/978-3-642-38762-3_3,

1 Introduction

For a correct description of reactive flow in porous media, the transport of the reactive species needs to be described correctly. As reaction takes place only in contact zones of the species involved, effective reaction rates are dependent on the microscopic concentration fields which can be strongly heterogeneous. The main problem in predicting concentration fields during tracer transport in a natural porous medium, such as soil, arises from the large range of scales involved. They span from the pore (micro) scale to the field (macro) scale, thus reaching from the range of or smaller than a μm to the km range. Thus different techniques are used to simulate tracer transport on different scales.

Transport in porous media is governed by three processes, the advective transport by the macroscopic flow field, the molecular diffusion and the mechanical dispersion due to the randomness of the individual streamlines through the pore space. Modelling dispersion on the macro scale has often been done by assuming an effective diffusivity for the tracer [1, 4]. The resulting advection-diffusion equation can be solved by standard discretisation methods (e.g. FE, FV and FD) or by stochastic (random walk) methods [9, 31]. Classical (FE, FV and FD) methods lack stability in advection dominated problems of tracer transport in porous media. Due to sharp gradients and front evolving in the solution, classical non-diffusive tend to produce non-physical oscillations. A way to get rid those oscillations is the introduction of numerical diffusion by upwinding. Another way is to stabilise the FE method by a variational multi-scale formulation [19, 38].

Modelling the mechanical dispersion by an effective diffusivity needs to regard two aspects, the non-Fickian regime in the initial phase and the dependence of the effective diffusivity on the randomness and structure of the porous matrix (e.g. soil). Special methods have been proposed to model non-Fickian dispersion in the initial phase by [9, 14, 20]. Such methods require knowledge of multi-point/multi-time statistics of the tracer and are therefore difficult to handle. On the other hand, the formulation of effective diffusion coefficients in the Fickian regime also requires knowledge on the randomness of the porous matrix. Preferential paths strongly amplify mechanical dispersion because in relatively slow regions, tracers can stay for a long time. Many studies therefore deal with the description of the permeability fields and their impact on tracer transport (e.g. [9]). In many cases, the parameters are subject to large uncertainties and can, if at all, only be described stochastically.

Recently, interest has grown in methods relying on velocity probability density functions (PDF). Meyer et al. have proposed a joint velocity-concentration PDF equation which accounts for advective transport and pore-scale dispersion in porous media and is solved by a particle method which is able to deal with non-Gaussian distributions of the velocity field [24, 25]; Jenny et al. introduce a new PDF method for obtaining information about tracer and phase transport by assuming that the multi-point velocity statistics is known [18]. Nowak et al. show the dependence of hydraulic heads and velocities on the variance of log-conductivity using Monte

Carlo simulations. They offer insight into the credibility of first-order second moment methods for evaluating moments of hydraulic heads. They observe a large deviation of the discharge components from Gaussian distribution and suggest using more accurate methods such as Monte Carlo if no assumptions on the shape of distributions are justified [26]. Deurer et al. [10] measured velocity PDFs in sphere packs by magnetic resonance imaging in various sample volumes to investigate longitudinal and transverse dispersion. They observed a strong dependence of the PDFs from sample volume.

In a research initiative on reactive flows in porous media, three different directions have been followed to improve prediction of concentration fields during the simulation of species transport through a porous medium. Our contributions are in the following fields: (i) proper resolution of the gradients of tracers without numerical diffusion on the macro scale (Sect. 5) (ii) description of subfilter fluctuations on micro-scale (Sect. 3) description of subfilter fluctuations on macro scale (Sect. 4).

The paper is organised as follows. In the next section, the equations describing flow in porous media, both on micro- as on macro-scale are discussed. After that, examples are presented that attack some of the problems in solving these equations by numerical methods. First, pore scale simulations using full solution of the Navier-Stokes equations are presented. Then, a method for parameter identification of an inhomogeneous permeability field is presented. Finally, a high order numerical method for transport on the macro-scale (Darcy-scale) is presented and discussed.

2 Description of Flow in Porous Media from Micro to Macro Scale

In this section some basic quantities on flow in porous media are defined. We start from a definition of the flow quantities on micro- and macro-scale as well as a discussion of the relevant equations of flow and tracer transport. The macro-scale equations are obtained by consequent homogenisation of the micro-scale equations over a representative elementary volume (REV). From this homogenisation, unclosed terms arise that have to be modelled adequately. Some problems of modelling and numerical solution of the respective equations are discussed.

We are considering incompressible flow of a Newtonian fluid and tracer transport through a porous medium. On the micro-scale, i.e. on volumes as large as the individual pores, the flow is governed by the Navier-Stokes equations, the conservation of mass

$$\nabla \cdot \mathbf{u} = 0 \tag{1}$$

and the conservation of momentum

$$\rho \partial_t \mathbf{u} + \rho \mathbf{u} \cdot \nabla \mathbf{u} = -\nabla p + \mu \nabla^2 \mathbf{u} \tag{2}$$

Here, $\mathbf{u}$, p, ρ and μ denote the velocity, pressure, density and dynamic viscosity, respectively. The transport of a tracer in the pore space is described by the convection diffusion equation for the concentration c of the tracer

$$\partial_t c + \mathbf{u} \cdot \nabla c = \Gamma \nabla^2 c \,. \tag{3}$$

Here, Γ is the molecular diffusivity.

The formalism of volume averaging [36] establishes a rigorous way of deriving macroscopic equations from the microscopic ones. If the total control volume, including fluid and solid phase, is denoted by V, then a superficial average of a quantity ψ can be defined the following way

$$\langle \psi \rangle = \frac{1}{V} \int_V \psi dx \,. \tag{4}$$

The porosity $\epsilon = V_p / V$ is defined to be the ratio of fluid filled volume (pore space V_p) divided by the total volume V. By volume-averaging the momentum equation (2) the well-known Darcy equation can be obtained

$$\langle \mathbf{u} \rangle = -\mathbf{K} \nabla \langle p \rangle \,, \tag{5}$$

in which $\mathbf{K}$ denotes the permeability tensor. However, when applying the averaging procedure on a larger scale, the definition of an *effective* permeability tensor poses problems as it is not a mere averaging of the permeability tensor at smaller scales. This can be seen by integrating equation (5) over a larger volume which gives

$$\langle \langle \mathbf{u} \rangle \rangle = - \langle \mathbf{K} \nabla \langle p \rangle \rangle \neq -\langle \mathbf{K} \rangle \nabla \langle p \rangle \,. \tag{6}$$

In measurements, often only large scale permeabilities are accessible, treated as effective permeabilities K_{eff}. If small scale variability of the permeability was accessible, the effective permeability can be obtained by up-scaling methods [12]

$$- \langle \mathbf{K} \nabla \langle p \rangle \rangle = -\mathbf{K}_{\text{eff}} \nabla \langle p \rangle \,. \tag{7}$$

The dispersion on a macro-scale is dependent on the distribution of the permeabilities on the scale of an REV as this determines whether e.g. preferential flow paths can establish.

When homogenising the convection diffusion equation (3), a similar problem arises. Averaging over an REV gives

$$\partial_t \langle c \rangle + \langle \mathbf{u} \cdot \nabla c \rangle = \Gamma \langle \nabla^2 \mathbf{c} \rangle \,. \tag{8}$$

In here, we have to realise that the second term on the left hand side causes problems, as $\langle \mathbf{u} \cdot \nabla c \rangle \neq \langle \mathbf{u} \rangle \cdot \nabla \langle c \rangle$. The underlying phenomenon is called dispersion. In most cases, it can be modelled by an additional diffusion using an effective dispersion coefficient [1]

$$\langle \mathbf{u} \cdot \nabla c \rangle = \langle \mathbf{u} \rangle \cdot \nabla \langle c \rangle + \Gamma^{\mathrm{disp}} \nabla^2 \langle c \rangle \, . \tag{9}$$

An effective dispersion is a good and valuable approach for late phases of tracer transport which are characterised by Gaussian tracer plumes [9]. Using $\Gamma^{\mathrm{eff}} = \Gamma + \Gamma^{\mathrm{disp}}$, Eq. (9) is then formulated as

$$\partial_t \langle c \rangle + \langle \mathbf{u} \rangle \cdot \nabla \langle c \rangle = \Gamma^{\mathrm{eff}} \nabla^2 \langle c \rangle \, . \tag{10}$$

These late stages are characterised by Fickian dispersion [9]. For early phases, strongly non-Gaussian tracer plumes and break-through curves are observed. These stages are characterised by non-Fickian dispersion and need special methods for description.

When flow and transport problems on a macro-scale are addressed the corresponding macroscopic parameters have to be modelled adequately, namely the effective permeability K_{eff} and the effective dispersion coefficient Γ^{eff}. Both can not be directly determined from basic principles. Either empirical correlations, experiments or numerical simulations on the micro-scale have to be used to estimate those macro-scale parameters.

In the following, we present some numerical efforts to improve our understanding of macro-scale parameters and processes. The first one addresses the description of dispersion by knowledge of the micro-scale velocity field, the second one deals with the estimation of the effective permeability distributions by macro-scale measurements and the third effort deals with the solution of the convection diffusion equation in convection dominated transport.

3 Pore Scale Simulations of the Flow Through a Random Sphere Pack

The variability of flow paths and velocities in porous media results in a dispersion of a tracer during its transport through a porous medium. Understanding the variability in the flow field is the key to understand and model dispersion in a rigorous way. The late phases of dispersion can be modelled by Fickian diffusion with an effective dispersion coefficient, see Eq. (10). Early phases, i.e. non-Fickian transport, need special attention as Eq. (10) can not represent non-Fickian behaviour which is often characterised by strongly skew break-through curves. In the following we present an attempt to understand flow variability in the pore space of a random sphere pack by describing the velocity distribution within the pore space.

We investigate the flow field on the pore scale of regular and random sphere packs by direct numerical simulation. The full Navier-Stokes equations (1) and (2) for an incompressible, Newtonian fluid are solved by a Finite Volume method on a Cartesian grid [23]. The irregular pore space is represented by an Immersed Boundary Method (IBM) to interpolate the no-slip boundary condition on the spheres to the Cartesian mesh [27, 28]. The spheres are represented by a triangular surface grid of triangle size smaller than the grid spacing of the Cartesian grid. The time advancement is done by a low-storage third order Runge-Kutta method [37]. This basic solver is well validated in various flow configurations including laminar and turbulent flows (e.g. [7, 17, 27]). It has been shown that for viscous flow problems a second order convergence with grid refinement is achieved [27, 28]. The sphere pack is generated by a special algorithm that distributes the spheres randomly in space. To achieve a periodic placement of the spheres, we first arranged the spheres on the faces of the domain. The inner part of the domain is then packed with as many spheres as possible. This method unfortunately results in a rather more porous area between the faces of the domain and the inner region that has to be taken into account in the post-processing.

We apply periodic boundary conditions in all three space dimensions. The flow is driven by a constant pressure gradient that is applied as a source term in the momentum equation. The simulation is advanced from rest until convergence has been reached. As the Reynolds numbers are extremely small, the time to reach convergence is mainly determined by the diffusion time scale within the pore space.

3.1 Grid Study

We checked the accuracy of the method by a convergence study of the flow through a regular sphere pack. In order to obtain the porous geometry we placed 23 spheres in a hexagonal packing arrangement and took out the smallest sized box that would fit into this arrangement and would be periodic in all three directions as our domain. We simulated low Reynolds number flow through this domain which was of size $(L_x, L_y, L_z) = (4, 2\sqrt{3}, 2\sqrt{3})$ mm. The flow was driven by a pressure gradient of $0.002\,\mathrm{P_a/m}$ in the x-direction. The Reynolds number of this setup was in the order of $Re = U_i D/\nu = 1 \times 10^{-5}$. Here, D is a characteristic length scale such as pore size or sphere diameter, and ν is the kinematic viscosity. U_i is the intrinsic velocity, which is defined as the mean pore velocity in the porous domain

$$U_i = \frac{1}{V_p} \int_{V_p} u(\mathbf{x})\, dv, \tag{11}$$

with V_p being the volume of the pore space. The intrinsic velocity is related to the superficial or Darcy velocity $\langle u \rangle$ by $U_i = \langle u \rangle / \epsilon$, where ϵ is the porosity.

We investigated the number of cells needed per sphere diameter for the bulk velocity to converge. Figure 1a shows the intrinsic velocity versus the number of grid cells per diameter of the grains and Fig. 1b shows the logarithm of the error (ε) in the computed intrinsic velocity as a function of the logarithm of grid cells per

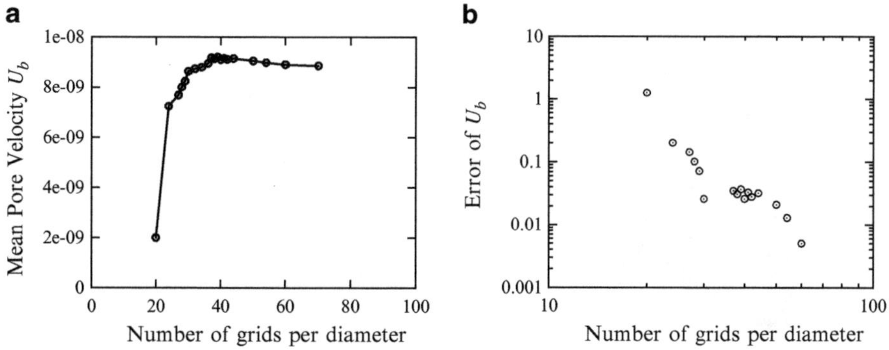

Fig. 1 (**a**) Mean pore velocity through a dense sphere pack as a function of number of grid cells per sphere diameter D. (**b**) Error of mean pore velocity as a function of grid cells per diameter D

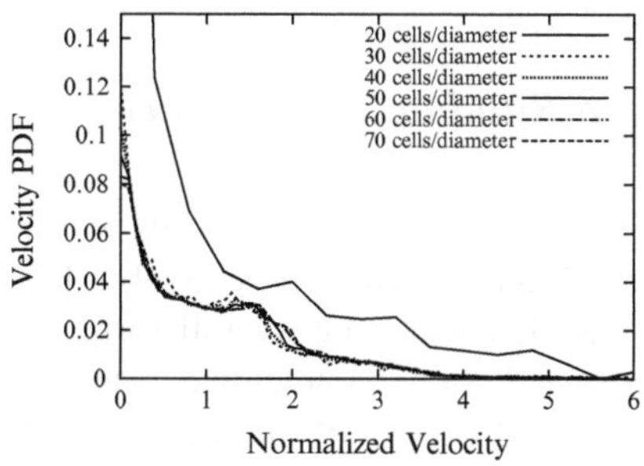

Fig. 2 Velocity PDFs in a dense sphere pack for various number of grid cells per sphere diameter D

diameter, taking the intrinsic velocity calculated using 70 grid cells per diameter as reference,

$$\varepsilon = \frac{U_i - U_{i,ref}}{U_{i,ref}}. \tag{12}$$

The intrinsic velocity converges monotonically with more than 35 cells per diameter and the error is limited to less than 5 %. There is no constant convergence rate due to the IBM method. However, on average, the convergence rate is at least of second order (Fig. 1b).

We inspect as well the probability density function (PDF) of the local velocity in pore space at different grid resolutions (Fig. 2). With more than 30 grid cells per diameter, the PDFs show only little variation. We concluded that with 40 grid cells per sphere diameter it will be possible to get a sufficiently accurate velocity field and chose such a mesh for the simulations presented in this work.

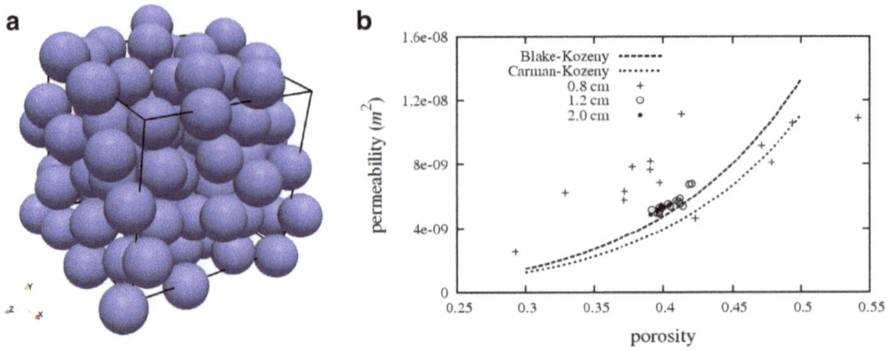

Fig. 3 (**a**) Random sphere pack: one realisation; (**b**) Comparison of computed permeabilites in the inner domain of random sphere pack domains of different sizes with the Blake-Kozeny and Carman-Kozeny relations

3.2 Flow Through a Random Sphere Pack

We did further validation by comparing our results to the empirical correlations on Carman-Kozeny and Blake-Kozeny, respectively. Those relations make use of dimensional analysis to determine the overall form of the dependence of the permeability from porosity and grain diameter in a sphere pack, Eq. (13). The factor α in this relation is related to the ratio of the mean length of the passages a flow has to go through and the thickness of the layer that it goes through and is fitted to experimental measurements. Carman-Kozeny is connected to $\alpha = 180$ while Blake-Kozeny is connected to $\alpha = 150$,

$$K = \frac{D^2 \epsilon^3}{\alpha(1-\epsilon)^2} . \tag{13}$$

A series of simulations through a random sphere pack with periodic boundary conditions in all three directions was conducted to find the minimum size of the REV. The grid resolution was 40 cells per diameter. The size of the domain increased from $0.8\,\text{cm} = 4D$ to $2\,\text{cm} = 10D$. For each domain size, we simulated 15 different realisations of random sphere distributions, such as displayed in Fig. 3a, to obtain a reasonable sample size. By this series, we can check which domain size can be regarded as REV. We found that close to the domain boundaries our porosity was little larger than in the inner domain where it was distributed homogeneously. Therefore, we take only the values from the inner domain for comparison with the Blake-Kozeny relation. This inspection revealed that a domain size of $10D$ was sufficient to obtain in the inner region permeability values fully consistent with Blake-Kozeny's relation, see Fig. 3b.

For every domain size we calculated the probability distribution function (PDF) of velocities in the range of $-2.6 \times 10^{-7}\,\text{m/s}$ and $8 \times 10^{-7}\,\text{m/s}$ using 1,325

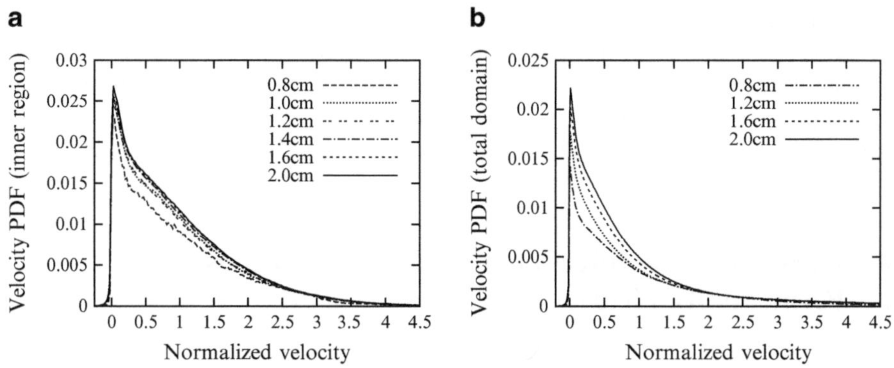

Fig. 4 Velocity PDF in a random sphere pack at various domain sizes. (**a**) inner domain only; (**b**) total domain

bins of size 8×10^{-10} m/s for each realisation, and then averaged the PDF over all 15 realisations. Because of the varying porosity in our domains due to the special sphere packing procedure mentioned before, we first calculated the PDF of velocities of points residing in the 'inner region' of the domain only. That is to say, in each direction we omitted the points closer than 1.5 sphere diameters to the edge and then proceeded to calculate the PDF of the velocities as mentioned before. In the next step we calculated these PDFs for the complete domain too. The PDFs of the stream-wise velocity in random sphere packs is plotted in Fig. 4a for the inner domain and Fig. 4b for the total domain. These plots demonstrate the convergence of the PDF with domain size. Here, the curves for the inner domain converge faster than the ones for the total domain which can be explained by the inhomogeneous porosity distribution close to the boundaries. The velocity in these plots is normalised by the intrinsic velocity U_i which is the average velocity in the pore space. The distribution is highly skew. Maximum velocities of four times the averaged one can be observed, however with very small likelihood.

Comparing our PDFs with those measured by magnetic resonance imaging [10], we observe large differences. Those were measured on various sample volumes, the smallest being in the range of sphere diameter. They represent velocities filtered on that scale. The maxima are at the order of magnitude of the pore velocity. Our PDFs have been evaluated at a sample size comparable to the grid spacing of the simulation which is much smaller than the sphere diameter. They can be regarded as unfiltered velocities and their maximum probability lies at values much smaller than the average pore velocity.

A striking feature of the PDFs are the negative velocities. Such negative velocities would not be expected in PDFs of the superficial velocities. They can be explained by the irregularity of the random sphere pack. This irregularity forces stagnation points at the front and back faces of the sphere to be off-centre. As a consequence, streamlines that travel to and from the stagnation points along the surfaces of the spheres have to point in negative x-direction in some regions and therefore generate negative stream-wise velocities. We conclude that those negative velocities can not be associated with flow separation in the traditional sense. Furthermore, we conclude that these negative velocities are not able to transport mass upstream over a long distance. However, they might increase the time a tracer needs to travel downstream and thus contribute to long tails of break-through curves.

4 Parameter Identification of an Inhomogeneous Permeability Field

In this section we focus on modeling flows in porous media on the macro-scale by the Darcy equation (5). One key problem is to determine the averaged material properties, here in particular the permeability tensors of the considered medium. Our approach is to determine them based on reference flow measurements taken from either experiments or direct numerical simulation resolving the micro-scale behaviour of the media. Then the permeability tensors are chosen such that the resulting flow given by the Darcy equation for the experiment configuration matches the measurements optimally in a least-squares sense. Previous work on parameter estimation in similar settings includes [22, 30] and [35].

We outline an adjoint-based optimisation algorithm that performs the parameter fit for a suitable discretisation of the Darcy model. Special emphasis is on a discretisation for the problem which on the one hand satisfies the necessary stability properties and on the other hand works well in the optimisation context. Tests on some model configurations show the viability of the proposed method.

Our model for describing a fluid moving through a porous domain $\Omega \subseteq \mathbb{R}^d$ consists of the Darcy equation (5) together with a volume integrated version of the mass balance equation (1). After rearranging the Darcy equation, it reads

$$\mathbf{K}_{\text{eff}}^{-1}\langle\mathbf{u}\rangle + \nabla\langle p\rangle = 0, \tag{14a}$$

$$\nabla\cdot\langle\mathbf{u}\rangle = f_p. \tag{14b}$$

The right hand side term f_p is used to model sources and sinks within the domain. By the position-dependent permeability tensor $\mathbf{K}_{\text{eff}}\colon \Omega \to \mathbb{R}^{d\times d}$ we describe the effective permeability of the media at any given point in the domain. We use a tensor instead of a scalar quantity since not only isotropic but also anisotropic materials should be modelled. According to [21], the tensor $\mathbf{K}_{\text{eff}}$ is symmetric positive definite at any given point in Ω. For our test configurations we assume homogeneous Neumann boundary conditions and the condition $\int_\Omega f_p\,\mathrm{d}x = 0$ which ensures existence and uniqueness of solutions for suitably chosen spaces for velocity, pressure and the permeability tensor.

Due to the saddle point structure of (14a), a finite element approximation has to be inf-sup stable. Since in the optimal control context we have to deal not only with the finite element spaces for the state variables but also with the corresponding dual spaces, using different Ansatz spaces for pressure and velocity would add considerable complexity. Therefore we want to use the same discrete spaces for both, pressure and velocity. Hence the inf-sup condition has to be enforced by a suitable stabilisation. Here we use the local projection stabilisation (LPS) approach (see [6]) since compared to most other methods the resulting stabilisation terms are symmetric. Therefore the two approaches *"discretise-then-optimise"* and *"optimise-then-discretise"* lead to the same set of discrete equations. In addition the systematic a posteriori error estimation approach developed in [2] can be applied immediately. For a detailed discussion of LPS stabilisation for optimal control, see [5]. A LPS stabilised discretisation of the Darcy-Brinkman has been analysed in [6]. Their results include the Darcy equation with homogeneous isotropic media as a special case and can be extended in a straight-forward fashion towards non-homogeneous anisotropic media. We use bi-linear rectangular finite elements on a conforming grid which possesses a patch structure, that is, the grid can be obtained by uniform refinement of a coarser grid $\mathcal{M}_h$. Then the stabilised discretisation of (14a) reads in weak form: find the discrete velocity and pressure $(\langle \mathbf{u}\rangle_h, \langle p\rangle_h)$ which satisfy

$$\begin{aligned}
&\int_\Omega \left\{\mathbf{K}_{\text{eff}}^{-1}\langle \mathbf{u}\rangle_h \varphi_v - \langle p\rangle_h \nabla\cdot\varphi_v + \varphi_p \nabla\cdot\langle \mathbf{u}\rangle_h\right\} \mathrm{d}x \\
&\quad + \sum_{M\in\mathcal{M}_h} \int_M \left\{h_M^2 \kappa_M(\nabla\langle \mathbf{u}\rangle_h)\kappa_M(\nabla\varphi_v) + \kappa_M(\nabla\langle p\rangle_h)\kappa_M(\nabla\varphi_p)\right\} \mathrm{d}x \\
&= \int_\Omega f_p \varphi_p \,\mathrm{d}x
\end{aligned}$$

for all discrete test functions (φ_v, φ_p). The fluctuation operator κ_M is defined locally on each cell $M \in \mathcal{M}_h$ of the coarser grid as $\kappa_M = \mathrm{Id} - \Pi_M$ with Id denoting the identity and Π_M the L^2 projection onto the space of constant functions on M. The diameter of M is denoted by h_M. Stability and first order convergence in the L^2 norm with respect to the discretisation parameter h are shown in [15].

For the parameter estimation problem we assume that we have a priori information about the distribution of different materials within the domain, furthermore that the domain can be divided into finitely many sharply bounded regions with different materials and that within each region the effective permeability tensor stays constant. In order to avoid enforcing the positive definiteness of the permeability tensor by additional constraints, we parametrise $\mathbf{K}_{\text{eff}}^{-1}$ in a suitable way by a finite number of parameters $q_i \in \mathbb{R}$. If we restrict our considerations to materials with diagonal permeability tensor, then a possible parametrisation consists of the d diagonal entries of $\mathbf{K}_{\text{eff}}^{-1}$ on each region. To ensure positive definiteness, the vector of parameters q is bounded away from zero by algebraic constraints.

Computing $\langle \mathbf{u}\rangle$ and $\langle p\rangle$ given $\mathbf{K}_{\text{eff}}^{-1}(q)$ is a well-posed problem, however the inverse problem of determining q from given measurements of $\langle \mathbf{u}\rangle$ and $\langle p\rangle$ can be ill-posed, that is, small variations in the measurement data can lead to big variations in the recovered q. Therefore we apply a Tychonoff regularisation with parameter $\alpha \geq 0$ such that the parameter identification problem can be stated as: Minimise

$$J(q,u) = \|Cu - z\|^2 + \frac{\alpha}{2}|q|^2 \tag{15}$$

subject to $u = (\langle \mathbf{u} \rangle, \langle p \rangle)$ solving the Darcy equation (14a) for $\mathbf{K}_{\text{eff}}^{-1} := \mathbf{K}_{\text{eff}}^{-1}(q)$ and $q \in Q_{\text{ad}} \subseteq Q = \mathbb{R}^N$ where N is the number of parameters in the parametrisation of the permeability tensor. The linear operator C models some measurements done on the computed solution, this could be for example evaluation of the velocity field at certain points within the domain. The value z represents the corresponding reference data for that measurement obtained from a micro-scale model or from an experiment. Since from a micro-scale simulation in principle we can obtain a complete reference state, it makes sense to chose the identity as observation operator C in that case. The parameter identification problem can be interpreted as an optimal control problem where the control variable q should be chosen in such a way that the state variable u matches a desired state described by the measurements as good as possible. We enforce positive definiteness of the permeability tensor by an appropriate choice of the closed set $Q_{\text{ad}} \subseteq Q$.

The existence of a solution to the optimal control problem can be shown by standard arguments, see for example the textbook [34]. Since the problem is in general non-convex, uniqueness of the solution cannot be guaranteed without further assumptions.

As noted before, for any control q there is a unique state u satisfying (14a). Therefore we can define the control-to-state mapping

$$S: q \mapsto u$$

with $u = (\langle \mathbf{u} \rangle, \langle p \rangle)$ solving the Darcy equation (14a) for $\mathbf{K}_{\text{eff}}^{-1} := \mathbf{K}_{\text{eff}}^{-1}(q)$. We introduce the reduced cost functional $j(q) := J(q, S(q))$ and state the reduced optimisation problem

$$\min j(q) \text{ subject to } q \in Q_{\text{ad}}.$$

To solve this reduced problem we use a primal-dual-active-set strategy (PDAS) (see, e. g., [16]) to treat the algebraic constraints on q resulting from the choice of Q_{ad}. In each step of the PDAS, an unconstrained optimisation problem has to be solved. For that purpose a globalised Newton-CG method is used. Gradient and Hessian information are computed via an adjoint approach, for further details on the algorithm see, e. g. [3] or [35]. To ensure fast convergence of the Newton method, exact derivatives that are consistent with the discrete stabilised state equation are essential. Therefore in particular the derivatives of the stabilisation terms with respect to q have to be taken into account when deriving the auxiliary equations used for Hessian evaluation.

Considering the computational complexity of the outlined algorithm, we note that the number of Newton steps does not depend on the fineness of the discretisation. The inner CG solver takes in the worst case $\mathcal{O}(N)$ iterations and for each iteration we have to solve two auxiliary PDEs, which each take $\mathcal{O}(L)$ operations with a multigrid solver, where L is the dimension of the finite element space. So in total we expect our algorithm to have the complexity $\mathcal{O}(N \cdot L)$.

For the numerical tests we consider the Darcy problem on the two-dimensional unit square $\Omega = (0,1)^2$. We subdivide Ω into 16 equally sized squares Ω_i, $i = 1, \dots, 16$ and assume that on each square the permeability tensor is constant and can be represented by a diagonal matrix. Therefore we choose the control space $Q = \mathbb{R}^{32}$ and define the parametrisation of the permeability tensor by

$$\mathbf{K}_{\text{eff}}^{-1}(q)|_{\Omega_i} = \begin{pmatrix} q_{2i-1} & 0 \\ 0 & q_{2i} \end{pmatrix} \quad \text{for } i = 1, \dots, 16.$$

For convenience we denote the vector collecting all the entries in the first component of $\mathbf{K}_{\text{eff}}^{-1}$ by $q^A \in \mathbb{R}^{16}$ and the one collecting the entries in the second component by q^B The source term is chosen as

$$f_p(x,y) = 2\cos(\pi x)\cos(\pi y),$$

and the set of admissible controls is defined as

$$Q_{\text{ad}} = \left\{ q \in \mathbb{R}^{32} | q \geq 1 \right\}.$$

Since the problem is reasonably well conditioned, we can omit the regularisation term by setting $\alpha = 0$. For the discretisation of pressure and velocity, a grid with 4,096 cells is used. The measurement data z are generated synthetically by performing a forward simulation with a reference parameter vector q_{ref}. We investigate two choices for the observation operator C, first the identity and second an operator modelling 32 point measurements of pressure and velocity within the domain. A visual comparison of the reference permeability tensor and the permeability tensors computed by the parameter identification algorithm can be seen in Fig. 5. For both choices of the observation operator C, good qualitative agreement between the reference and the computed permeability values is observed. However, for the case $C = \text{Id}$, the estimated parameters are better than for the point-wise measurements since more data enters the computation. These observations are confirmed when looking at the relative errors $\frac{\|q^A - q^A_{\text{ref}}\|_2}{\|q^A_{\text{ref}}\|_2}$ and $\frac{\|q^B - q^B_{\text{ref}}\|_2}{\|q^B_{\text{ref}}\|_2}$ listed in Table 1. A qualitative comparison of the resulting velocity fields to the reference velocity field is shown in Fig. 6.

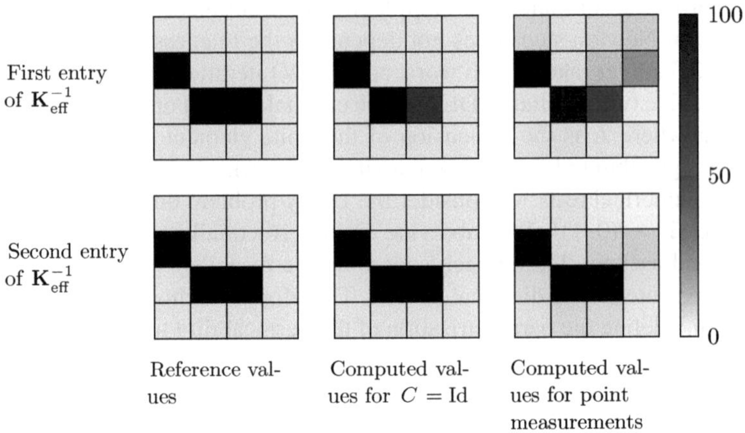

Fig. 5 Values of $\mathbf{K}_{\text{eff}}^{-1}$ over the domain Ω for $C = \text{Id}$ and for 32 point measurements

Table 1 Relative errors of the two tensor components for both choices of C

	$C = \text{Id}$	Point measurements
$\frac{\lVert q^A - q^A_{\text{ref}} \rVert_2}{\lVert q^A_{\text{ref}} \rVert_2}$	0.0655	0.181
$\frac{\lVert q^B - q^B_{\text{ref}} \rVert_2}{\lVert q^B_{\text{ref}} \rVert_2}$	0.00565	0.0634

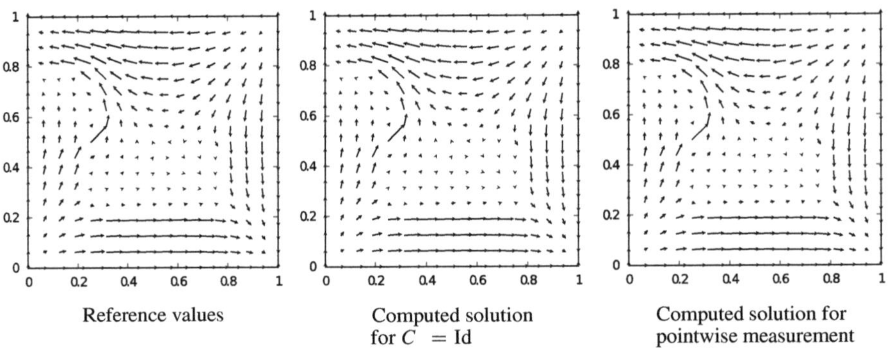

Fig. 6 Comparison of exact velocity field and velocity fields resulting from estimated q

5 High Order Finite Element Method for the Advection Diffusion Equation

One of the main numerical problems for simulations of tracer transport on the macro-scale are strong gradients within the tracer fields developing in situations were convection dominates over diffusion. Standard Bubnov-Galerkin finite elements are known to deliver oscillating solutions for convection dominated problems for meshes which are not fine enough. It has not yet been proved whether raising the polynomial degree of the shape functions will increase or decrease numerical oscillations. This paper will show that an increase of the polynomial degree (p-FEM) stabilises the numerical oscillations in Bubnov-Galerkin type finite elements naturally without adding any additional stabilisation term.

We will demonstrate the improvement of the numerical accuracy with polynomial order using a one-dimensional stationary convection-diffusion problem (10). Given a constant convection velocity u_x, a steady and constant effective diffusion coefficient Γ^{eff} and a source term f, the problem is to find $c : \Omega \to \mathbb{R}$, such that with Dirichlet boundary conditions

$$\begin{cases} u_x \dfrac{dc}{dx} - \Gamma^{\text{eff}} \dfrac{d^2 c}{dx^2} = f & \text{on} \quad \Omega = \{x | 0 < x < 1\} \\ c = 0 & \text{at} \quad x = 0 \\ c = 0 & \text{at} \quad x = 1 \end{cases} \tag{16}$$

We contrast the numerical errors of p-FEM [33] to the standard h-FEM [32] in which linear shape functions are used and follow the analysis scheme presented in [11]. Herein, the truncation error of a Bubnov-Galerkin discretisation is quantified in order to specify the additional diffusion term used in Petrov-Galerkin methods. For a h-FEM, this results in a discretised equation which includes the numerical diffusion $\bar{\Gamma}^{\text{eff}}$

$$u_x \left(\frac{c_{j+1} - c_{j-1}}{2h} \right) - (\Gamma^{\text{eff}} + \bar{\Gamma}^{\text{eff}}) \left(\frac{c_{j+1} - 2c_j + c_{j-1}}{h^2} \right) = 1\,. \tag{17}$$

The extra term $\bar{\Gamma}^{\text{eff}}$ can be interpreted either as the truncation error of the Bubnov-Galerkin method of first order or as an additional diffusivity required to provide nodally exact results. This term is a function of the mesh *Péclet* number and reads

$$\bar{\Gamma}^{\text{eff}} = \left(\coth Pe - \frac{1}{Pe} \right) \Gamma^{\text{eff}} Pe\,. \tag{18}$$

The mesh *Péclet* number is defined as

$$Pe = \frac{u_x h}{2\Gamma^{\text{eff}}}\,. \tag{19}$$

where h is the mesh or grid size.

The value of $\bar{\Gamma}^{\text{eff}}$ increases with the mesh *Péclet* number. In fact, Eq. (18) forms the basic motivation behind using the Petrov-Galerkin method. In many stabilisation approaches, one tries to control the artificial numerical oscillations in convection dominated problems by compensating for the truncation error by means of adding additional diffusivity. However, it will be shown in the next section that the truncation error of the Bubnov-Galerkin method is decreased by a mere increase of the polynomial order of the spatial discretisation.

It is important to mention here that the truncation error study shown in next sections is also performed in more details in [8]. In the paper [8], the stabilization capability of the p-FEM for convection-dominated transport problems is explained mathematically by analyzing stiffness matrices. Numerical examples show that using sufficiently high order polynomial degrees for shape functions can eliminate the nodal oscillations in numerical solutions for convection-dominated problems, where the mesh *Péclet* number is greater than one. This approach will be introduced in following sections again in order to explain, why the high order FEM is suitable for solving convection-dominated problems of tracer transport on the macro-scale.

5.1 Truncation Error of the Bubnov-Galerkin Discretisation in the p-FEM

In this section, the truncation error of p-FEM for the same example as presented above is considered, where hierarchic shape functions of second order derived from the set of integrated Legendre polynomials are applied and the polynomial orders up to 5 are investigated. Compared to Lagrange shape functions, hierarchic shape functions are easy to construct since lower order shape functions are subsets of higher order ones. We refer to [32], where the complete hierarchy of spaces is introduced.

In general, the system equation using polynomial degrees higher than 2 can be also condensed analogously as in Eq. (17), using $\bar{\Gamma}^{\text{eff}}{}_p$ instead of $\bar{\Gamma}^{\text{eff}}$ as all higher modes are purely internal to the element.

Analogous to the previous analysis, one can get the following diffusion using second to fifth order polynomials for shape functions, respectively.

$$
\begin{aligned}
\bar{\Gamma}^{\text{eff}}{}_2 &= \frac{1}{3} Pe^2 \Gamma^{\text{eff}} \\
\bar{\Gamma}^{\text{eff}}{}_3 &= \frac{5Pe^2 \Gamma^{\text{eff}}}{Pe^2 + 15} \\
\bar{\Gamma}^{\text{eff}}{}_4 &= \frac{\Gamma^{\text{eff}}(Pe^4 + 35Pe^2)}{10Pe^2 + 105} \\
\bar{\Gamma}^{\text{eff}}{}_5 &= \frac{14\Gamma^{\text{eff}}(4Pe^4 + 90Pe^2)}{4Pe^4 + 420Pe^2 + 3780}
\end{aligned}
\tag{20}
$$

The truncation error of p-FEM is defined as

$$
\Delta\Gamma_p^{\text{eff}} = \bar{\Gamma}^{\text{eff}} - \bar{\Gamma}^{\text{eff}}{}_p \tag{21}
$$

and depicted in dependence of Pe in Fig. 7, where the ordinate displays $\Delta\Gamma_p^{\text{eff}}$.

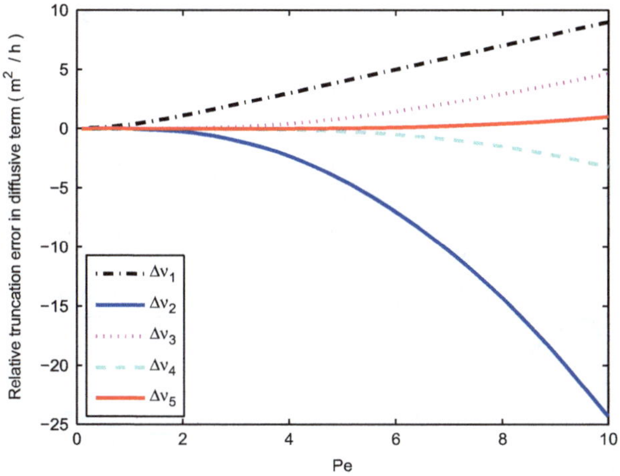

Fig. 7 Truncation error with different polynomial degrees

In general, the curves have different tendencies which correspond to the parity of the polynomial degree. Odd degrees generate curves which increase monotonically as *Pe* increases while the even ones decrease. Although the sign of truncation error depends on the parity of the order, the absolute value of truncation error decreases when the order of shape functions grows. Accordingly, the numerical solution at nodes approaches the exact solution.

On the other hand, using odd polynomial degrees, the numerical diffusivity of the high order approach is less than $\bar{\Gamma}^{\text{eff}}$. This lack of diffusivity is the reason of the oscillatory behaviour of the numerical solution at high *Pe*. By contrast, the numerical diffusivity is always greater than $\bar{\Gamma}^{\text{eff}}$ using even polynomial degrees. Consequently, nodal solutions exhibit an over-diffusive behaviour and never show nodal oscillations. This result is further analysed from a mathematical perspective in the next section.

5.2 *Connection of the Stability and the Structure of the System Matrix*

Stability, i.e. oscillations or not, is determined by the structure of the system matrix. The numerical simulation will start to oscillate if the discrete maximum principle is violated [29]. Considering a system matrix structure such as given in Eq. (22), it can be proved that no oscillations occur for $\alpha < 1$ [13].

$$\mathscr{A}(\alpha) = tridiag(-1-\alpha, 2, -1+\alpha) \tag{22}$$

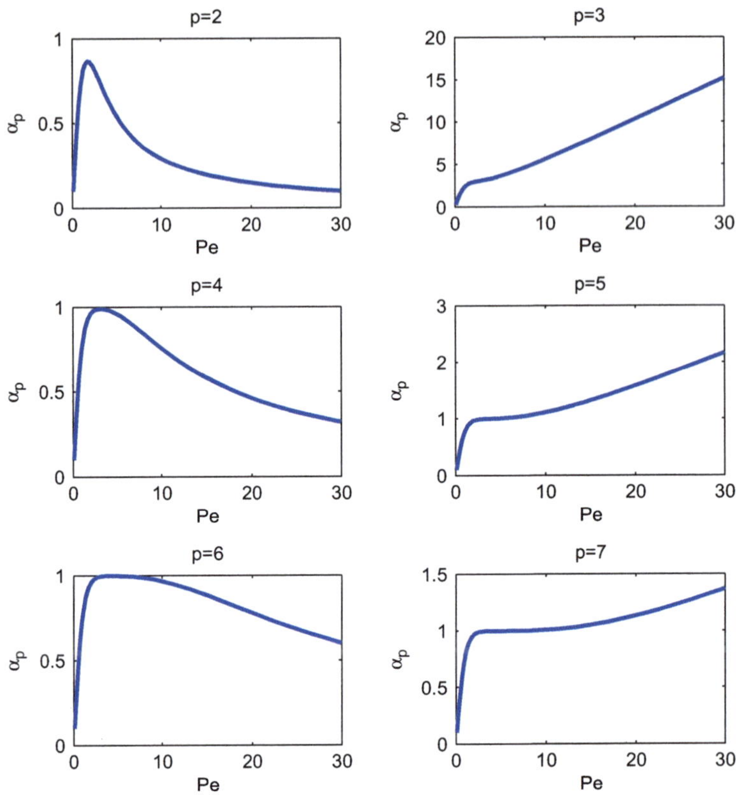

Fig. 8 α_p behaves differently for odd and even polynomial degrees

The system matrix resulting from the condensed equation (17) can be written as

$$\mathscr{A}_p = \frac{(\Gamma^{\mathrm{eff}} + \bar{\Gamma}^{\mathrm{eff}}{}_p)}{h^2} tridiag(-1 - \alpha_p, 2, -1 + \alpha_p)\,, \qquad \alpha_p = \frac{u_x h}{2(\Gamma^{\mathrm{eff}} + \bar{\Gamma}^{\mathrm{eff}}{}_p)}\,. \tag{23}$$

Consequently, the stability of nodal solutions is determined by the value of α_p. Further, the value of α_p can be quantified for higher order polynomial degrees based on Eq. (20):

The corresponding values are plotted in Fig. 8. It can be observed that for odd polynomial degrees α_p increases as *Pe*. For even polynomial degrees, α_p first increases and then decreases while the value is always smaller than 1. This in turn means that for even polynomial degrees, the numerical solution at nodal degrees of freedom never oscillates. This result also coincides with the conclusion from the truncation error analysis in the previous section. To further clarify this point, we plot the solution of the 1D example with $Pe = 20$ shown in Fig. 9.

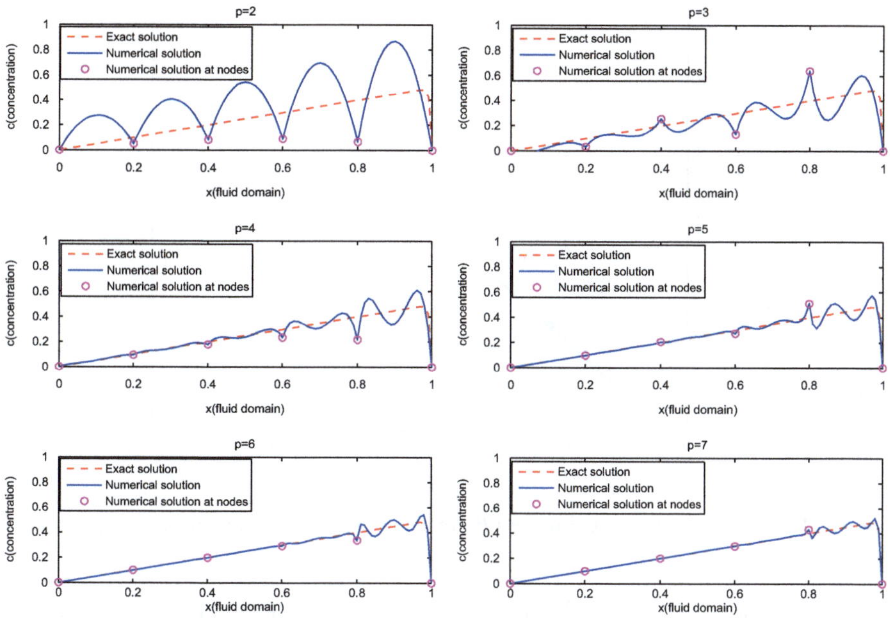

Fig. 9 Comparison of numerical, exact and nodal solutions with different Ansatz degree, Pe = 20

Here, the exact solution denotes the analytical solution of the differential equation (16). Figure 9 illustrates that when the polynomial degree is even, numerical oscillations only stem from internal modes and numerical solutions at each node do not oscillate. For odd polynomial degrees, numerical oscillations are reflected by both internal and nodal degrees of freedom.

By setting $\alpha_p = 1$ in Eq. (23), we can compute the maximum allowed *Pe* which guarantees nodally stable solutions for the given polynomial degree of the shape functions. In other words, for a given mesh *Péclet* number, the corresponding p stated in Eq. (24) is the minimum required polynomial degree and their relationship is depicted in Fig. 10. It turns out to be almost linear for polynomial orders $p \leq 11$.

$$\begin{aligned} p &= 3 \quad Pe = 2.322185 \\ p &= 5 \quad Pe = 3.646738 \\ p &= 7 \quad Pe = 4.971786 \\ p &= 9 \quad Pe = 6.297019 \\ p &= 11 \quad Pe = 7.622340 \\ &\ldots \end{aligned} \tag{24}$$

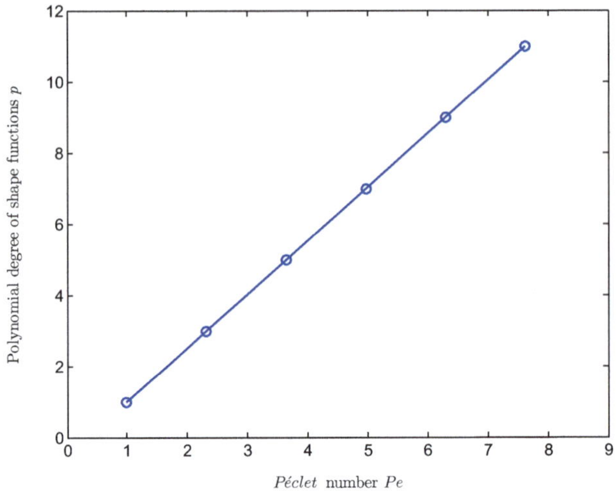

Fig. 10 The relation between a given mesh *Péclet* number and the minimum required polynomial degree

Different from other up-winding methods, where the additional efforts for modelling the necessary artificial diffusivity for more complicated problems, the high order FEM generates the additional numerical diffusion naturally by purely increasing the polynomial degrees. In the following example, numerical results of the one-dimensional convection-diffusion transport problem are compared to the exact solution. The given differential equation (25)

$$\begin{cases} a\dfrac{dc}{dx} - \Gamma^{\text{eff}}\dfrac{d^2c}{dx^2} = 0 & \text{on} \quad \Omega = \{x|0 < x < 1\} \\ c = 0 & \text{at} \quad x = 0 \\ c = 1 & \text{at} \quad x = 1 \end{cases} \tag{25}$$

has the analytical solution

$$y = \frac{e^{ax/\Gamma^{\text{eff}}} - 1}{e^{a/\Gamma^{\text{eff}}} - 1}. \tag{26}$$

When the mesh is fixed, the ratio between a velocity and a diffusivity determines the mesh *Péclet* number and characterises the convergence of the numerical solution. When the mesh *Péclet* number increases, the standard Bubnov-Galerkin method based on linear elements exhibits oscillations in the numerical solution. We choose the parameters $a = 2\,\text{m/h}$, $\Gamma^{\text{eff}} = 0.02\,\text{m}^2/\text{h}$, and compute the corresponding numerical solutions with 10 elements of the same length $h = 0.1$. Figure 11 shows numerical solutions with different polynomial degrees. The dashed line denotes the exact solution while the solid line represents the numerical solution.

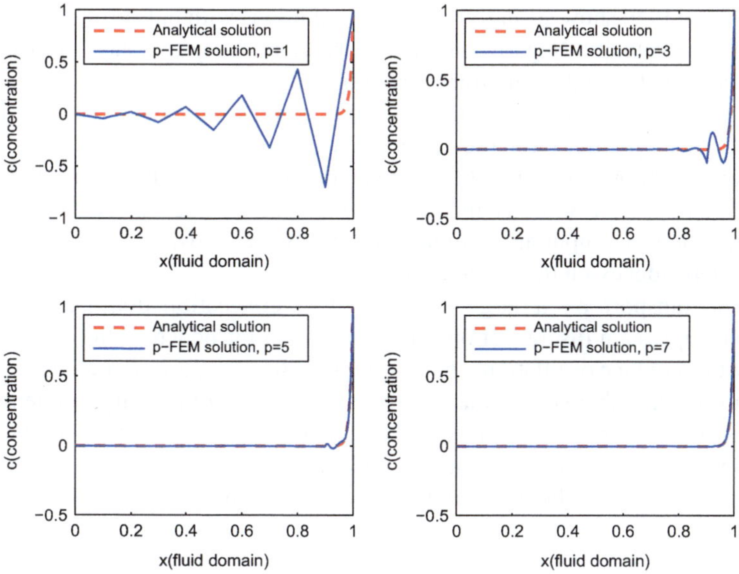

Fig. 11 Numerical solution with different polynomial degrees, $Pe = 5$

As expected, when mesh *Péclet* number $Pe = \dfrac{ah}{2\Gamma^{\mathrm{eff}}} = 5$ is larger than 1, the numerical solution with linear Bubnov-Galerkin discretisation introduces non-physical oscillations. The p-FEM can eliminate these oscillations by simply raising the polynomial degree p. It is observed in Fig. 11 that with $p = 7$, the oscillation is drastically suppressed and the numerical solution is in good agreement with the analytical one.

6 Conclusions

In this paper we presented some efforts to improve understanding and simulation of flow and transport in porous media. Using consistent volume averaging, it can be shown that traditional closures, such as effective permeability and diffusivity are not applicable in all situations. Those situations arise for dispersion in the initial phase of tracer transport, for strongly inhomogeneous permeability fields and for convection dominated transport.

The initial phase of tracer transport is characterised by non-Gaussian tracer plumes, the so-called non-Fickian regime. The transition from non-Fickian to Fickian dispersion is dependent on how long tracer patches stay in low-speed regions. To understand this phenomenon, we investigated the PDF of the stream-wise velocity by detailed simulations of the flow in the pore space of random sphere packs. These PDF show strongly skewed distributions with tails up to four times the average pore velocity. Negative velocities are more likely to delay tracer transport than to contribute to upstream transport of tracer.

The determination of the inhomogeneous permeability field can contribute to understand and predict the large-scale tracer dispersion. We presented an adjoint-based optimisation algorithm to estimate permeability distributions from point measurements of the velocity in a porous medium. The results show a satisfying agreement between input and estimated permeability fields. As expected, they also reveal a dependency on the observation operator.

Tracer transport on a large scale is often convection dominated. In these situations, upstream discretisations are used which introduce additional numerical diffusivity to reduce oscillations in the solution. However, this numerical diffusivity is not always a viable solution as it strongly smears out the sharp gradients in the tracer field. In this paper, we presented a numerical analysis of the p-FEM method to determine under which conditions unphysical oscillations can be damped by the use of higher order methods without introducing unwanted numerical diffusion.

References

1. Bear, J.: Dynamics of Fluids in Porous Media. American Elsevier Publishing Company, New York (1972)
2. Becker, R., Vexler, B.: A posteriori error estimation for finite element discretization of parameter identification problems. Numer. Math. **96**(3), 435–459 (2004)
3. Becker, R., Meidner, D., Vexler, B.: Efficient numerical solution of parabolic optimization problems by finite element methods. Optim. Methods Softw. **22**(5), 813–833 (2007)
4. Borges da Silva, E., Souza, D., Ulson de Souza, A., Guelli U. de Souza, S.: Prediction of effective diffusivity tensors for bulk diffusion with chemical reactions in porous media. Braz. J. Chem. Eng. **24**, 47–60 (2007)
5. Braack, M.: Optimal control in fluid mechanics by finite elements with symmetric stabilization. SIAM J. Control Optim. **48**(2), 672–687 (2009)
6. Braack, M., Schieweck, F.: Equal-order finite elements with local projection stabilization for the darcy-brinkman equations. Comp. Methods Appl. Mech. Eng. **200**(9–12), 1126–1136 (2011)
7. Breuer, M., Peller, N., Rapp, C., Manhart, M.: Flow over periodic hills – numerical and experimental study over a wide range of Reynolds numbers. Comput. Fluids **38**, 433–457 (2009)
8. Cai, Q., Kollmannsberger, S., Sala, E., Huerta, A., Rank, E.: On the natural stabilization of convection dominated problems using high order Bubnov-Galerkin finite elements. Comput. Math. Appl. (2013, submitted)
9. Dentz, M., Cortis, A., Scher, H., Berkowitz, B.: Time behavior of solute transport in heterogeneous media: transition from anomalous to normal transport. Adv. Water Resour. **27**, 155–173 (2004)
10. Deurer, M., Vogeler, I., Clothier, B., Scotter, D.: Magnetic resonance imaging of hydrodynamic dispersion in a saturated porous medium. Transp. Porous Media **54**, 145–166 (2004)
11. Donea, J., Huerta, A.: Finite Element Methods for Flow Problems. Wiley, Chichester/Hoboken (2003)

12. Durlofsky, L.: Numerical calculation of equivalent grid block permeability tensors for heterogeneous porous media. Water Resour. Res. **27**(5), 699–708 (1991)
13. Ernst, O.: Residual-minimizing Krylov subspace methods for stabilized discretization of convection-diffusion equations. SIAM J. Matrix Anal. Appl. **21**, 1079–1101 (2000)
14. Hassanizadeh, S.: On the transient non-Fickian dispersion theory. Transp. Porous Media **23**, 107–124 (1996)
15. Himmelstoß, T.: Stabilisierung und Parameteridentifikation für die Darcy-Gleichungen. Master's thesis, Technische Universität München (2011)
16. Hintermüller, M., Ito, K., Kunisch, K.: The primal-dual active set strategy as a semismooth Newton method. SIAM J. Optim. **13**(3), 865–888 (2003)
17. Hokpunna, A., Manhart, M.: Compact fourth-order finite volume method for numerical solutions of Navier-Stokes equations on staggered grids. J. Comput. Phys. **229**, 7545–7570 (2010)
18. Jenny, P., Tchelepi, H., Meyer, D.: Uncertainty assessment of transport in porous media based on a probability density function method. In: Proceedings of 10th European Conference on the Mathematics of Oil Recovery, Amsterdam (2006)
19. Juanes, R.: A variational multiscale finite element method for multiphase flow in porous media. Finite Elem. Anal. Des. **41**(7–8), 763–777 (2005)
20. Levy, M., Berkowitz, B.: Measurement and analysis of non-Fickian dispersion in heterogeneous porous media. J. Contam. Hydrol. **64**, 203–226 (2003)
21. Liakopoulos, A.: Darcy's coefficient of permeability as symmetric tensor of second rank. Int. Assoc. Sci. Hydrol. Bull. **10**(3), 41–48 (1965)
22. Mahnken, R., Steinmann, P.: A finite element algorithm for parameter identification of material models for fluid-saturated porous media. Int. J. Numer. Anal. Methods Geomech. **25**(5), 415–434 (2001)
23. Manhart, M.: A zonal grid algorithm for DNS of turbulent boundary layers. Comput. Fluids **33**(3), 435–461 (2004)
24. Meyer, D., Jenny, P.: Stochastic mixing model for PDF simulations of turbulent reacting flow. In: Advances in Turbulence X, pp. 681–684. CIMNE, Barcelona (2004)
25. Meyer, D., Tchelepi, H., Jenny, P.: An eulerian joint velocity-concentration PDF method for solute dispersion in highly heterogeneous porous media. Geophys. Res. Abstr. **12**, 5700 (2010)
26. Nowak, W., Schwede, R., Cirpca, O., Neuweiler, I.: Probability density functions of hydraulic head and velocity in three-dimensional heterogeneous porous media. Water Resour. Res. **44** (2008)
27. Peller, N.: Numerische simulation turbulenter Strömungen mit immersed boundaries. Ph.D. thesis, Technische Universität München (2010)
28. Peller, N., Le Duc, A., Tremblay, F., Manhart, M.: High-order stable interpolations for immersed boundary methods. Int. J. Numer. Methods Fluids **52**, 1175–1193 (2006)
29. Rank, E., Katz, C., Werner, H.: On the importance of the discrete maximum principle in transient analysis using finite element methods. Int. J. Numer. Methods Eng. **19**, 1771–1782 (1983)
30. Schulz, V., Wittum, G.: Multigrid optimization methods for stationary parameter identification problems in groundwater flow. In: Multigrid Methods V. Volume 3 of the LCSE series, pp. 276–288. Springer, Berlin (1998)
31. Suciu, N., Radu, F., Prechtel, A., Knabner, P.: A coupled finite element-global random walk approach to advection-dominated transport in porous media with random hydraulic conductivity. Comput. Appl. Math. **246**, 27–37 (2013)
32. Szabó, B., Babuška, I.: Finite Element Analysis. Wiley, New York (1991)
33. Szabó, B., Düster, A., Rank, E.: The p-version of the finite element method. Encycl. Comput. Mech. **1**, 119–139 (2004)
34. Tröltzsch, F.: Optimal Control of Partial Differential Equations: Theory, Methods and Applications. American Mathematical Society, Providence (2010)
35. Vexler, B.: Adaptive finite element methods for parameter identification problems. Ph.D. thesis, Universität Heidelberg, Naturwissenschaftlich-Mathematische Gesamtfakultät (2004)

36. Whitaker, S.: Flow in porous media I: a theoretical derivation of Darcy's law. Transp. Porous Media **1**(1), 3–25 (1986)
37. Williamson, J.: Low-storage Runge-Kutta schemes. J. Comput. Phys. **35**(48) (1980)
38. Yang, C., Samper, J.: A subgrid-scale stabilized finite element method for multicomponent reactive transport through porous media. Transp. Porous Media **78**(1), 101–126 (2009)

Optimal Control of Partially Miscible Two-Phase Flow with Applications to Subsurface CO_2 Sequestration

Moritz Simon and Michael Ulbrich

Abstract Motivated by applications in subsurface CO_2 sequestration, we investigate constrained optimal control problems with partially miscible two-phase flow in porous media. The objective is, e.g., to maximize the amount of trapped CO_2 in an underground reservoir after a fixed period of CO_2 injection, where the time-dependent injection rates in multiple wells are used as control parameters. We describe the governing two-phase two-component Darcy flow PDE system and formulate the optimal control problem. For the discretization we use a variant of the BOX method, a locally conservative control-volume FE method. The timestep-wise Lagrangian of the control problem is implemented as a functional in the PDE toolbox Sundance, which is part of the HPC software Trilinos. The resulting MPI parallelized Sundance state and adjoint solvers are linked to the interior point optimization package IPOPT. Finally, we present some numerical results in a heterogeneous model reservoir.

Keywords Partially miscible two-phase flow • Optimal control • Adjoint approach • Complementarity condition • Control-volume FE method • Variational formulation • CO_2 sequestration

1 Introduction

In this paper, we present an adjoint based [24] approach for optimal control problems that are governed by multiphase multicomponent flow in porous media. The concrete application motivating our work is optimal CO_2 sequestration in underground reservoirs: The goal is to maximize the amount of trapped CO_2 after a

M. Simon (✉) · M. Ulbrich
Chair of Mathematical Optimization, Department of Mathematics, Technische Universität München, Boltzmannstr. 3, 85748 Garching b. München, Germany
e-mail: simon@ma.tum.de; mulbrich@ma.tum.de

M. Bader et al. (eds.), *Advanced Computing*, Lecture Notes in Computational Science and Engineering 93, DOI 10.1007/978-3-642-38762-3_4,

fixed time interval of injection into a 2D reservoir Ω. To this end, we use a partially miscible two-phase two-component flow model [7]. The discretization (currently in 2D) is done with the aid of the PDE toolbox Sundance [30], which is part of the Trilinos framework [22]. We extended Sundance by the finite volume (FV) and upwinding capabilities required to implement the BOX method [3] and developed an interface between Sundance and the interior point optimization package IPOPT [42]. Extensions to 3D and more complex flows are possible.

For the discretization of the state system PDEs we apply a variant of the BOX method [3], a locally conservative control-volume FE method. For further information on this method see [21, 25]. Other possible numerical methods include streamline diffusion techniques [45] and conservative mixed FE methods [35]. For a survey of suitable FE/FV methods for multiphase multicomponent flows in porous media we refer to [14] and references therein.

Concerning related work on PDE-constrained optimization in reservoir modeling, the main focus has so far not been on CO_2 trapping mechanisms, but rather on inverse history matching, i.e., parameter estimation in petroleum reservoirs (see [34] for an overview), or on the economic aspect to maximize oil production. To this end, optimal control of mixed CO_2–water injection was first considered in [32], using IMPES (implicit-pressure–explicit-saturation) simulation of a modified black oil model and gradient methods for optimization. For recent approaches to optimal control of enhanced oil recovery (EOR) via water-flooding we refer to the surveys [11, 26] and to the original articles [2, 12, 15, 36, 38, 39, 41, 43].

The optimal control problem we consider here is the optimal sequestration of CO_2 in underground reservoirs. To obtain a suitable two-phase model, we restrict our attention to the phases CO_2 and water/brine, neglecting oil and further substances. This leads to a partially miscible two-phase flow model. Note that the mentioned control approaches to optimal water-flooding usually apply immiscible two-phase or black oil type three-phase flow models. Thus, in our context, the miscibility of CO_2 in brine introduces a further degree of complexity in the state system. We model the switching between saturation and undersaturation of the wetting phase with CO_2 by a complementarity condition, as also suggested in [4, 27].

Concerning our software choices, we intended to stay as flexible as possible. We work with Sundance, which among other advantages allows for a direct variational problem formulation [31] that, in particular, supports the direct derivation of the exact discrete adjoint. To achieve this, the timestep-wise Lagrangian of the control problem is formulated as a Sundance functional. We have implemented an interface to the interior point software IPOPT [42], a state-of-the-art nonlinear optimization code that can handle general constraints and provides L-BFGS [46] approximations for the Hessian of the Lagrangian. In [18] several test problems are considered to show that limited-memory BFGS techniques are superior to other comparable algorithms when it comes to EOR history matching with multiphase flow.

The paper is organized as follows: Sect. 2 introduces the state equations for Darcy based partially miscible two-phase flow, together with primary state variables and control objectives. Section 3 presents the adjoint based optimal control framework, while the applied numerical methods for the state equation are sketched in Sect. 4.

Implementation strategies for the adjoint based framework within the PDE toolbox Sundance will be discussed in Sect. 5. Finally, Sect. 6 contains a discussion of optimization results, and some conclusions are presented in Sect. 7.

2 State Equations, Variables and Control Objectives

The CO_2 injection takes place at wells in a subregion $\Omega_c \subset \Omega$ of the underground reservoir Ω, being purely saturated with saline water (called "brine") at the beginning. The thermodynamical reservoir conditions are chosen such that the injected CO_2 is in supercritical state, so we consider a continuous underground flow model with two phases $\alpha \in \{w, n\}$ and two components $i \in \{1, 2\}$. This two-phase flow is assumed partially miscible in the following sense: On the one hand, the CO_2 component $i = 2$ can (partially or fully) dissolve in the water-rich wetting phase $\alpha = w$. On the other hand, the "evaporation" of the water component $i = 1$ into the CO_2-rich nonwetting phase $\alpha = n$ is neglected.

The reservoir medium is assumed rigid with time-invariant functions $\phi = \phi(x)$ and $\mathbf{K} = \mathbf{K}(x)$ for porosity and absolute permeability. Moreover, we neglect temporal and spatial variations of the reservoir temperature and assume chemical equilibrium. The most important physical agreement that determines the flow model's nature is the application of Darcy's law in the PDE system. The mass conservation equations for the components $i \in \{1, 2\}$ ($1 =$ water, $2 = CO_2$) within the phases $\alpha \in \{w, n\}$ (wetting and nonwetting) [7] are then given by

$$\phi \partial_t \sum_\alpha \rho_\alpha X_\alpha^i S_\alpha - \sum_\alpha \operatorname{div}\Big[\lambda_\alpha \rho_\alpha X_\alpha^i \mathbf{K}(\nabla p_\alpha + \rho_\alpha \mathbf{g} e_2)\Big] - \operatorname{div}\big(D^i \rho_w \nabla X_w^i\big) = q^i. \quad (1)$$

Here ρ_α, S_α and p_α denote the phase densities, saturations and pressures, while X_α^i stands for the mass fraction of component i in phase α. Especially, we have $X_n^1 = 0$ and $X_n^2 = 1$ in the partially miscible setting, while we set $X := X_w^2$ so that $X_w^1 = 1 - X$. The first divergence term in (1) expresses advection and buoyancy (here $\mathbf{g}$ denotes gravitational acceleration) via Darcy's law with phase mobilities λ_α and absolute permeability matrix $\mathbf{K}$, whereas the second divergence term models the slow diffusion in the wetting phase. The source terms q^i are discussed below. Next, we explain some modeling agreements.

For phase mobilities λ_α and capillary pressure p_c we use Brooks-Corey models [10]; the phase viscosities are assumed constant. The brine phase is incompressible, but a linear density increase with the CO_2 mass fraction X is taken into account. Slight compressibility of the CO_2 phase is expressed in a linearized density $\rho_n(p_n)$. The nonlinear diffusion coefficient $D^2 = D^2(\phi, S_w)$ is taken from [7, Sect. 2.5], while water–brine diffusion is neglected via $D^1 = 0$. For further details on these models we refer to [37].

As we do not consider thermodynamical heat exchange, the governing PDE system is given by (1) for the components water and CO_2. However, the CO_2

component is allowed to dissolve in the wetting phase $\alpha = w$ to some extent. Hence we need three primary state variables $y := (p_n, S_w, X)$ in the flow model:

1. CO_2 pressure: $p_n = p_n(x,t) \longrightarrow p_w = p_n - p_c$ with $p_c = p_c(S_w)$,
2. Brine saturation: $S_w = S_w(x,t) \longrightarrow S_n = 1 - S_w$ (natural constraint),
3. CO_2 mass fraction: $X = X(x,t) \longrightarrow X = \varphi(p_n)$ when $S_w(x,t) < 1$.

As an initial condition we state $S_w(\cdot, 0) = 1$ (no CO_2 injected); in such single-phase regimes the pressure $p_n = p_w$ exists as a mutual parameter. Any value $X < \varphi(p_n)$ corresponds to $S_w = 1$, while the nonwetting phase $\alpha = n$ can only appear (that means $S_w < 1$) if X reaches the threshold $\varphi(p_n)$. In the presence of the nonwetting phase we have the relation $X = \varphi(p_n)$. Thus an adaptive switching between the variables X and S_w is required; we model this by a complementarity condition:

$$a := 1 - S_w \geq 0, \quad b := \varphi(p_n) - X \geq 0 \quad \text{and} \quad ab = 0.$$

This can equivalently be written as $\tilde{g}(a,b) = 0$, where $\tilde{g}$ is a complementarity function, e.g., the Fischer-Burmeister [17] function $\tilde{g}(a,b) = a + b - \sqrt{a^2 + b^2}$. To maintain differentiability, we smooth this nonsmooth equation in our implementation to obtain an algebraic equation $g(p_n, S_w, X) = 0$ that holds pointwise in space and time (see [37] for details). The use of complementarity conditions to model phase transitions in miscible multiphase flow has also been suggested in [4, 27].

With the above state variables $y = (p_n, S_w, X)$, connected through the complementarity condition $g(y) = 0$, the flow system (1) can be recast in a more compact form: Introducing the scalar and vector valued abbreviations

$$b^1 := \rho_w(1 - X)S_w, \quad b^2 := \rho_n(1 - S_w) + \rho_w X S_w,$$

$$B^1 := \lambda_w \rho_w (1 - X)\mathbf{K}\big[\nabla(p_n - p_c) + \rho_w \mathbf{g} e_2\big] - D^1 \rho_w \nabla X,$$

$$B^2 := \lambda_w \rho_w X \mathbf{K}\big[\nabla(p_n - p_c) + \rho_w \mathbf{g} e_2\big] + \lambda_n \rho_n \mathbf{K}\big(\nabla p_n + \rho_n \mathbf{g} e_2\big) + D^2 \rho_w \nabla X,$$

we obtain an algebro-differential system of the form

$$\phi \partial_t b^i(y) - \operatorname{div} B^i(y, \nabla y, x) = q^i(u) \quad (i = 1, 2) \quad \text{and} \quad g(y) = 0 \qquad (2)$$

in the space-time interior domain $\Omega \times (0, T)$. The spatial boundary $\partial\Omega = \Gamma_n \cup \Gamma_d$ is decomposed into Neumann and Dirichlet parts: A no-flux condition $n^T B^i(y, \nabla y, x) = 0$ for $i = 1, 2$ is considered on Γ_n (impermeable cap rock). The conditions $y = y_d$ on Γ_d impose standard values without CO_2 injection, also taken as initial values inside the brine-saturated reservoir at time $t = 0$.

The control parameters $u = u(t)$ only appear in the source terms $q^i = q^i(u)$:

$$q^1 = 0 \quad \text{and} \quad q^2 = \sum_{n=1}^{N} q_n(t)\omega(x - x_n).$$

This means no water is injected, while q^2 models the injection of supercritical CO_2 at $N \geq 1$ injection wells with locations x_n in a certain control area $\Omega_c \subset \Omega$; ω models the injection distribution, given, e.g., by a narrow Gaussian normal distribution. In the following, the N injection rates $q_n(t)$ are used as control parameters:

$$u(t) = \big(q_1(t), \ldots, q_N(t)\big), \quad t \in [0, T]. \tag{3}$$

We impose box constraints on the above rates and make sure that the total amount of injected CO_2 does not exceed a certain threshold. This leads to control constraints

$$0 \leq q_n(t) \leq a, \quad t \in [0, T], \quad 1 \leq n \leq N \quad \text{and} \quad \frac{1}{T}\sum_{n=1}^{N}\int_0^T q_n(t)\,\mathrm{d}t \leq b \tag{4}$$

for fixed numbers $a, b > 0$ with $b < Na$ (otherwise the latter constraint is redundant). For future investigation, it will be promising to also include additional state constraints, such as an upper bound on the pressure p_n to preserve caprock integrity. State constraints are known to be challenging from a theoretical point-of-view, see, e.g., [24, Sect. 2.7.2] and [40]. A suitable approach to tackle such constraints is by regularization (e.g., of Moreau-Yosida type) [24, 33, 40] and our optimization interface is capable of handling state constraints by a regularization approach. In this paper, however, we only consider control constraints.

Our objective is to maximize the amount of trapped CO_2 after a finite time $T > 0$ of injection. Here we consider the following types of "trapping": CO_2 can be

1. Immobilized below a certain residual saturation S_n^r (residual trapping),
2. Dissolved in the brine and thereby trapped for the moment (solubility trapping).

The residual saturation $S_n^r = S_n^r(\phi)$ depends on the porosity. Below this threshold, i.e., for saturations $S_n \leq S_n^r$, the nonwetting phase mobility λ_n drops to zero. For a discussion of further trapping objectives we refer to [37].

A general optimization objective concerns the final amount of trapped CO_2 in either of the above variants, weighted by coefficients $\beta_{1,2} \in [0, 1]$, which results in an objective functional of the form

$$J(y, u) = \int_\Omega \big(\beta_1 \phi \rho_n S_n \mathbf{1}_{\{S_n \leq S_n^r\}} + \beta_2 \phi \rho_w X S_w\big)\big|_{t=T}\,\mathrm{d}x - \gamma R(u), \tag{5}$$

where an additional control-dependent penalty term $R(u)$ allows for regularization of the control, where $\gamma \geq 0$ is a regularization parameter. It is chosen in a weighted H^1-type manner to penalize strong fluctuations in the wells' rates with an additional weighted L^2-type regularization. Explicitly, this strictly convex regularization is given by

$$R(u) := \frac{q_*^2}{T} \sum_{n=1}^{N} \int_0^T \left[T^2 q_n'(t)^2 + \theta(t) q_n(t)^2 \right] \mathrm{d}t, \tag{6}$$

including a factor q_* ($q_* = 200$ in SI units) and a time-dependent weighting function $\theta : [0, T] \to [0, \infty)$.

We finally remark that many concrete choices made in this section could be replaced by a wide range of other choices without leaving our overall framework.

3 Adjoint Based Optimal Control Framework

Provided that the weak state equation $E(y, u) = 0$ has a unique solution $y = y(u)$ for every control u and, moreover, that the operator $E(y, u)$ and the functional $J(y, u)$ are sufficiently smooth, the adjoint approach [24] can be used for efficient computation of the derivative of the reduced functional $j(u) := J\big(y(u), u\big)$:

$$j'(u) = E_u(y, u)^* \mu + J_u(y, u),$$

where the adjoint state μ solves the adjoint equation

$$E_y(y, u)^* \mu + J_y(y, u) = 0.$$

Here the subscripts u and y denote partial (Fréchet-)derivatives with respect to the variables u and y, e.g., $E_y(y, u) = \frac{\partial}{\partial y} E(y, u)$, and $E_y(y, u)^*$ is the adjoint operator of the linear operator $E_y(y, u)$.

In order to calculate $j(u)$ and $j'(u)$, discretizations of both state and adjoint equation must be set up (see [37] for the derivation of the continuous adjoint equation). The numerical methods we apply are implemented such that the discretized adjoint system coincides with the exact discrete adjoint of the discretized state equation ("discretize–then–optimize").

The time horizon $t \in [0, T]$ is discretized into n_t equidistant time points t_k with time step size $\Delta t := t_k - t_{k-1}$. The corresponding vector

$$u \in \mathbb{R}^{N n_t} \quad \text{with} \quad u_{(n-1)n_t+k} := q_* q_n(t_k), \quad 1 \le k \le n_t, \quad 1 \le n \le N, \tag{7}$$

discretizes the control variables (3) with the mentioned scaling factor q_*.

Currently, we do not use adaptive time stepping, but this could be integrated: In fact, whereas the spatial discretization and the computation of adjoints in space are obtained automatically from the capabilities of Sundance, time stepping for forward and adjoint solves is implemented manually as a for-loop and can be adjusted accordingly. It is known that adaptive time stepping requires care in order to ensure sufficient accuracy of the adjoint based gradient. In our context, a local (in terms of the control) freezing of the time grid during adjoint based derivative computations should be an appropriate way to include adaptivity in time [19,23,28]. For a selection of work on adjoint based optimal control with adaptive time stepping, we refer to [1, 16, 19, 23, 28].

The control constraints (4) in discrete form read

$$0 \le u_j \le q_* a, \quad 1 \le j \le Nn_t \quad \text{and} \quad c(u) := \frac{1}{n_t} \sum_{j=1}^{Nn_t} u_j \le q_* b,$$

while the penalty term (6) can be discretized as

$$R(u) \approx \sum_{n=1}^{N} \left\{ n_t \sum_{k=1}^{n_t - 1} \left[u_{(n-1)n_t+k+1} - u_{(n-1)n_t+k} \right]^2 + \frac{1}{n_t} \sum_{k=1}^{n_t} \theta_k u^2_{(n-1)n_t+k} \right\}. \quad (8)$$

Here we set $\theta_k := \theta(t_k)$ for the user-defined weighting coefficients.

To solve the resulting constrained nonconvex optimization problem, the interior point based software package IPOPT [42] was chosen. We developed a C++ interface that calls the Sundance state and adjoint solvers to compute the cost function $j(u)$ and its derivative $j'(u)$. The interior point algorithm in IPOPT is configured to use limited-memory BFGS updates [46] for approximating the Hessian matrix of the Lagrange function.

4 Numerical Methods for the State Equation

Denoting by y^k the time-discretized state variables at time $t = t_k$ and applying an implicit Euler scheme to the state system (2), we obtain the following semi-discretized state equation in strong form:

$$\phi \frac{b^i(y^k) - b^i(y^{k-1})}{\Delta t} - \operatorname{div} B^i\left(y^k, \nabla y^k, x\right) - q^i\big|_{t=t_k} = 0 \quad (i = 1, 2) \quad (9)$$

together with the algebraic complementarity $g(y^k) = 0$ and the boundary conditions for $1 \le k \le n_t$; the initial condition $y^0 = y_d$ reproduces the Dirichlet values.

Concerning the space discretization, we use the BOX method [3], a control-volume FE method. It has two main advantages: On the one hand it is locally conservative due to its control-volume nature, on the other hand it possesses sufficient stability due to full upwinding [20] and mass lumping [25]. Both the control-volume and upwinding features were integrated into the PDE toolbox Sundance [30] within the Trilinos framework to make it accessible for such stabilized locally conservative methods.

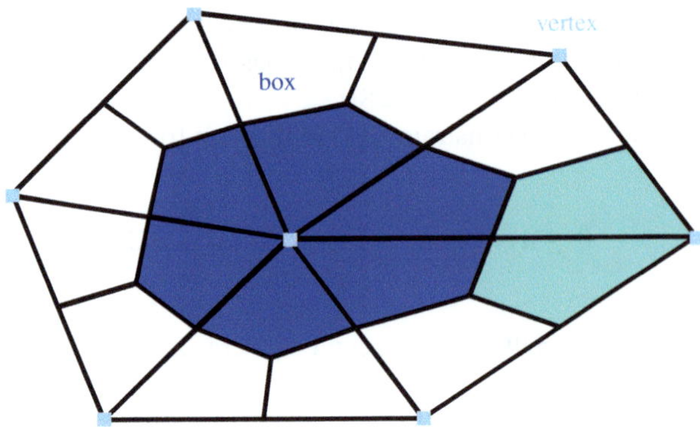

Fig. 1 Construction of a box in the dual mesh

The BOX method uses piecewise linear continuous finite elements on a primal (in our case triangular) mesh for the discretization of the state y^k. Similar to finite volume methods, constant functions on the vertex-centered cells ("boxes") of the dual mesh are used as test functions; see Fig. 1 for the construction of a box in the dual mesh of control volumes.

Given a primal triangulation of the domain $\Omega \subset \mathbb{R}^2$, we denote the collection of boxes in the dual mesh by B_j $(j = 1, \dots, m)$ and their respective box volumes by ω_j. To derive the space discretization, the semi-discrete state equation (9) is integrated over each box B_j – note that the indicator functions $\mathbf{1}_{B_j}$ are a basis of our test space – and the Gauß divergence theorem is applied to obtain

$$\int_{B_j} \left[\phi \frac{b^i(y^k) - b^i(y^{k-1})}{\Delta t} - \operatorname{div} B^i(y^k, \nabla y^k, x) - q^i\big|_{t=t_k} \right] \mathrm{d}x$$

$$= \int_{B_j} \left[\phi \frac{b^i(y^k) - b^i(y^{k-1})}{\Delta t} - q^i\big|_{t=t_k} \right] \mathrm{d}x - \int_{\partial B_j} n^T B^i(y^k, \nabla y^k, x)\, \mathrm{d}S = 0. \tag{10}$$

Inserting piecewise linear Lagrange finite elements for the state and applying upwinding to the term $n^T B^i$ in the boxwise weak form (10), we then arrive at

$$\int_{B_j} \left[\phi \frac{b^i(y^k) - b^i(y^{k-1})}{\Delta t} - q^i\big|_{t=t_k} \right] \mathrm{d}x - \int_{\partial B_j} n^T B^i_{\mathrm{up}}\, \mathrm{d}S = 0 \quad (i = 1, 2). \tag{11}$$

The upwinding procedure for the modified terms B^i_{up} in the boundary integrals of (11) will be explained below. Quadrature rules for the boxes are chosen as follows: Volume integrals are evaluated vertex-centered, i.e.,

$$\int_{B_j} f(x)\, \mathrm{d}x \approx \omega_j f(v_j), \quad 1 \le j \le m,$$

where v_j denotes the unique primal vertex inside the box B_j (see Fig. 1). For the underlying control-volume FE method this yields a mass lumping, where the entries of the FE mass matrix are assigned to its main diagonal [25]. The use of the lumped mass matrix counteracts the appearance of non-physical oscillations [13].

The boundary integrals $\int_{\partial B_j} n^T B^i_{\text{up}}\, dS$ are evaluated via a midpoint rule on each line segment of the polygonal curve ∂B_j in triangular 2D grids. The fully implicit upwinding procedure is based on the decomposition

$$B^i = -\sum_{\alpha} C^i_\alpha V_\alpha + D^i \rho_w X^i_w \quad \text{with} \quad C^i_\alpha := \lambda_\alpha \rho_\alpha X^i_\alpha, \quad V_\alpha := -\mathbf{K}(\nabla p_\alpha + \rho_\alpha \mathbf{g} e_2).$$

Upon implicit evaluation of the Darcy velocities V_α at the discretized state y^k, the above advective coefficients C^i_α are implicitly evaluated

- At the vertex v_j of the box B_j for outflow $n^T V_\alpha \geq 0$,
- At the vertex of the opposite box for inflow $n^T V_\alpha < 0$.

Here n denotes the outer dual cell normal of the box B_j and "opposite" in 2D is subject to the respective line segment of the polygonal curve ∂B_j (see Fig. 1). The evaluation of the upwinded flux terms B^i_{up} is based on this strategy. For further details see [3, Chap. 3] for two-phase flow or [7, Sect. 4.2] for miscible flow.

5 Implementation Strategies in Trilinos/Sundance

As a software platform for state and adjoint simulations, the choice was made for the FE library package Sundance [30], based on the scientific computing framework Trilinos [22]. Within the MAC-TUM project B7, Sandia National Labs, Texas Tech University and Technische Universität München (TUM) have joined forces to further develop and improve this toolbox, especially in view of PDE-constrained optimization and multiphysics problems [5, 6, 31]. The mentioned C++ interface of the Sundance simulation code to IPOPT is based on earlier work on shape optimization with Navier-Stokes flow [8, 9, 29].

Sundance provides a wide range of benefits, such as efficient treatment of nonlinearities, automatic generation of linearized equations and thereby support for implementing adjoints within efficient gradient based optimization algorithms [31]. In our context, we represent the whole optimal control problem by its Lagrange function $\mathscr{L}(y, u, \mu)$ as a timestep-wise functional in Sundance. Here μ denotes the adjoint state, i.e., the Lagrange multiplier of the state equation. Sundance conveniently provides routines that automatically derive nonlinear and linear variational problems from this corresponding to $\mathscr{L}_{\mu^k} = 0$, which is the state equation, and $\mathscr{L}_{y^k} = 0$, which is the adjoint equation. Furthermore, Sundance is MPI based and provides parallel system assembly as well as interfaces to a whole library of direct and iterative parallel solvers (e.g., NOX, Amesos, AztecOO).

In order to make these convenient features available for our purposes, we had to implement several additional class structures and quadrature rules within the PDE toolbox, since Sundance so far had not directly supported control volumes and upwinding, as explained in Sect. 4. Since those two issues are essential ingredients of the BOX method, we integrated a 2D implementation of these features into the object oriented C++ framework of Sundance.

An extension to 3D can be achieved with manageable effort. Before we proceed in describing our 2D implementation, we briefly discuss the question of extendibility to 3D: As illustrated below, our top-level coding of the problem is very flexible and thus, for 3D, would mainly require to adjust the Lagrange function of the optimal control problem to this setting. The structure of the code for state and adjoint computations and the interface to the optimization method is essentially independent of the space dimension. Of course, the problem dimension increases significantly in 3D and an efficient preconditioning of the Krylov subspace solvers would be necessary. The main work for extending our implementation to 3D would have to be spent on extending several geometric low-level subroutines (dual elements etc.) from 2D to 3D. For 3D quad grids this is possible with reasonable efforts, whereas unstructured tetrahedral meshes in 3D require more low-level implementation work. An alternative for extending our code to 3D would be to stack copies of the 2D grid to extrude it along the third dimension. This approach would shift the required implementation work from the low-level routines to the top-level problem coding.

In the following, we illustrate the mentioned advantages for the implementation of state and adjoint equation systems: The state equation for y^k and its adjoint for μ^k at time step t_k can be written in terms of the time-discretized Lagrangian:

$$\mathscr{L}_{\mu^k}(y,u,\mu) = 0 \quad \text{(time-discretized state equation)},$$
$$\mathscr{L}_{y^k}(y,u,\mu) = 0 \quad \text{(corresponding adjoint equation)},$$
$$\text{where} \quad y := \left(y^1,\dots,y^{n_t}\right), \quad \mu := \left(\mu^1,\dots,\mu^{n_t}\right).$$

This Lagrangian $\mathscr{L}$ contains the stepwise weak state equation functionals $\langle \mu^k, E_k \rangle$ and the stepwise evaluations of the objective functional (5) as follows:

$$\mathscr{L}(y,u,\mu) = \sum_{k=1}^{n_t} L_k\left(y^k, y^{k-1}, u, \mu^k\right) \quad \text{with timestep-wise Lagrangian}$$

$$L_k := \Delta t \left\langle \mu^k, E_k\left(y^k, y^{k-1}, u\right)\right\rangle + \delta_{kn_t} \int_\Omega f\left(y^{n_t}\right) \mathrm{d}x - \gamma r_k(u),$$

where f abbreviates the final-time density in the integral of (5) and the terms r_k denote the stepwise components of the time-discretized penalty term (8).

Since μ^k only appears in the functional L_k, whereas y^k appears both in L_k and L_{k+1} for $1 \leq k < n_t$ and only in L_{n_t} for $k = n_t$, this means that, defining two further functionals F and G at a respective fixed time step t_k via

$$F := L_k \quad \text{and} \quad G := L_k + (1 - \delta_{kn_t})L_{k+1},$$

one can write the weak time-discretized state and adjoint equations as $F_{\mu^k} = 0$ and $G_{y^k} = 0$. This provides the basis for our flexible implementation of the discrete state system on the one hand and its exact discrete adjoint system on the other hand.

Sundance allows to implement functionals in a very concise way. For example, the Sundance expression

```
Expr auv = Integral(Omega, (grad*u)*(grad*v), quad);
```

generates the bilinear form corresponding to $a(u, v) = \int_\Omega \nabla u \cdot \nabla v \, dx$, using the quadrature rule quad. Here, u and v are FE functions, e.g., given by

```
Expr u = new UnknownFunction(new Lagrange(1), "u");
```

as piecewise linear Lagrange elements. Forms representing Dirichlet boundary conditions can be defined with the method EssentialBC instead of Integral. The command

```
Functional L(mesh, intform, bcform, vecType);
```

combines the Integral form intform and the EssentialBC form bcform to construct a functional L. Here mesh describes the computational grid and vecType is the Trilinos vector type. Along these lines the functionals F and G can be implemented in Sundance as Functional F and G, respectively. For efficiency reasons we include only those parts of L_{k+1} in G that depend on the current state y^k. The methods

```
NonlinearProblem
F.nonlinearVariationalProb(mu,mus,y,ys,...);
LinearProblem
G.linearVariationalProb(y,ys,mu,...);
```

generate a nonlinear problem for the state time step equation $F_{\mu^k}(y^k) = 0$ and a respective linear problem for the adjoint time step equation $G_{y^k}(\mu^k) = 0$. Here y is the unknown vector of the state variables y^k (with several additional components to facilitate upwinding) and mu denotes a corresponding adjoint variable vector. The FE functions ys and mus provide storage for y and mu, while the dots hide further fixed parameters, for instance y^{k+1} and μ^{k+1} in $G_{y^k}(\mu^k) = 0$.

The order of parameters in nonlinearVariationalProb is: variable w.r.t. which the functional shall be linearized (here: mu); value of the linearization point (here: mus); unknown variable for which the resulting equation shall be solved (here: y); storage for previous variable (here: ys); fixed variables (here: not written out); values of the fixed variables (here: not written out). The order of parameters for linearVariationalProb applies accordingly, but no storage is provided for the unknown variable.

For runtime acceleration, we implemented an IMPES-like strategy (here IMPES stands for implicit-pressure–explicit-saturation) to generate good initial iterates for implicit time steps that turn out to require too many NOX iterations (see [37] for details). Furthermore, the state and adjoint simulation was parallelized by domain decomposition. The parallelization uses in-built features of Sundance with MPI based communication between processors.

6 Discussion of Optimization Results

SI units are suppressed during the discussion. The following case study was performed in a rectangular reservoir $\Omega = (0, a_1) \times (0, a_2)$ of height $a_2 = 100$ and width $a_1 = 4a_2$, having the Neumann boundary $\Gamma_n = (0, a_1)\times\{0, a_2\}$ on the top and bottom and the Dirichlet boundary $\Gamma_d = \{0, a_1\}\times(0, a_2)$ on the side strips. Dirichlet conditions are chosen as $S_w = 1$, $p_n = 8 \cdot 10^6$ on top (otherwise hydrostatic) and $X \approx 10^{-4}$, so that complementarity $g(y) = 0$ holds on this boundary.

Moreover, we restrict our attention to a reservoir with constant porosity $\phi = 0.2$ and impose constant residual saturations $S_w^r = S_n^r = 0.1$ for both water and CO_2. The Sundance simulation is done on a triangular grid with $n_1 \times n_2$ uniform rectangles in Ω (two triangles per rectangle). However, we mention that our implementation supports unstructured grids. In what follows, a horizontal domain decomposition with $n_p = 16$ parallel processes is applied on a mesh with $n_1 = 128$ and $n_2 = 32$.

The rectangular model reservoir is supplied with a heterogeneous isotropic absolute permeability distribution $\mathbf{K}(x) = k_0(x)\mathbf{id}$ and with $N = 5$ injection wells. The function $k_0(x)$ is illustrated in Fig. 2: The permeability increases from left to right and has a horizontal layer of significantly lowered permeability. The five wells are ordered in a trapezoidal pattern. Notice that the two upper wells are located within the blue layer of low permeability.

The optimization was run for $T = 4$ months with $n_t = 200$ time steps and has converged up to an IPOPT tolerance of $2 \cdot 10^{-5}$, in our case corresponding to a dual infeasibility norm (see [42] for precise definitions); the objective function was scaled by -10^{-3} in the interface, since IPOPT minimizes and expects problem-adjusted scaling. We set $\beta_1 = \beta_2 = 1$ in (5) for a combination of residual and solubility trapping in the objective, while the regularization parameter is $\gamma = 5 \cdot 10^{-4}$ and the weight function $\theta(t) = 4 \cdot 10^3 (1 + \frac{4t}{T})$ in (6) penalizes late injection. Given a scaling of $q_* = 200$ in (7), the control constraints

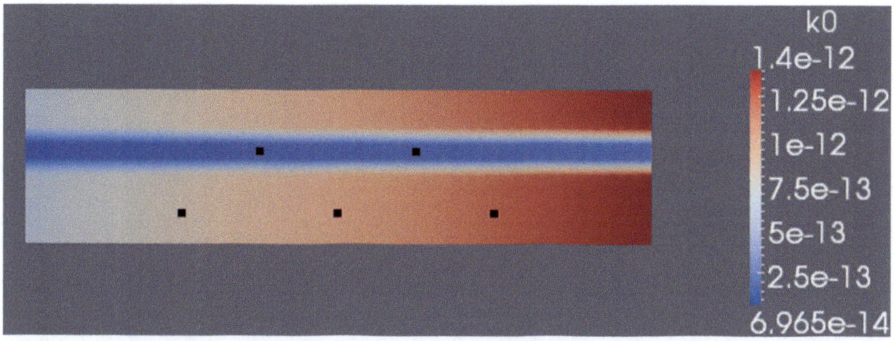

Fig. 2 Permeability distribution k_0 and injection well locations

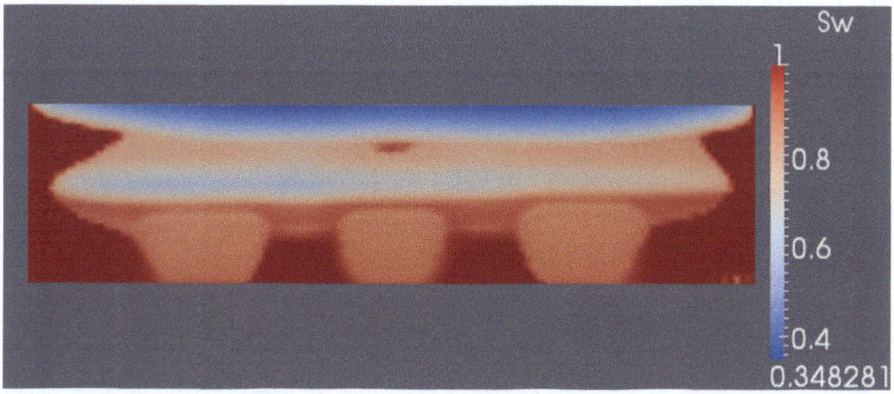

Fig. 3 Final saturation S_w for optimal control

$$0 \leq u_j \leq 9 \quad \text{and} \quad c(u) = \frac{1}{n_t} \sum_{j=1}^{Nn_t} u_j \leq 18$$

make sure that at least two wells must be active for maximal injection $c(u) = 18$.

Figures 3–5 illustrate the final saturation profiles for the computed optimal injection strategy in comparison to an early bang-bang strategy – i.e., inject the maximal amount in all wells for the first 80 time steps, then switch injection off – and to uniform injection with $c(u) = 18$. In contrast to other results in [37], the optimal control has $c(u) \approx 17.856 < 18$, so the upper bound is not active in this situation. This behavior shows that one must not inject too much CO_2 into the reservoir. The objective functional values are:

- Optimal control: $J = 2.465 \cdot 10^5$, $J\big|_{\gamma=0} = 2.471 \cdot 10^5$,
- Early bang-bang strategy: $J\big|_{\gamma=0} = 2.429 \cdot 10^5$,
- Uniform injection ($u_j = 3.6$): $J\big|_{\gamma=0} = 1.965 \cdot 10^5$.

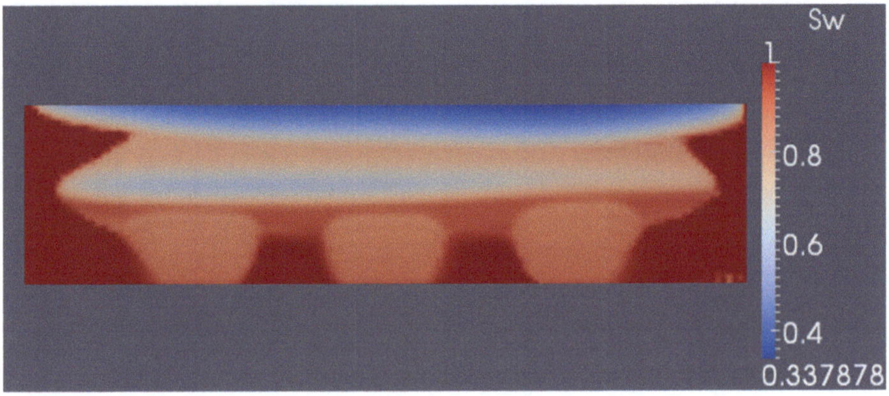

Fig. 4 Final saturation S_w for early bang-bang

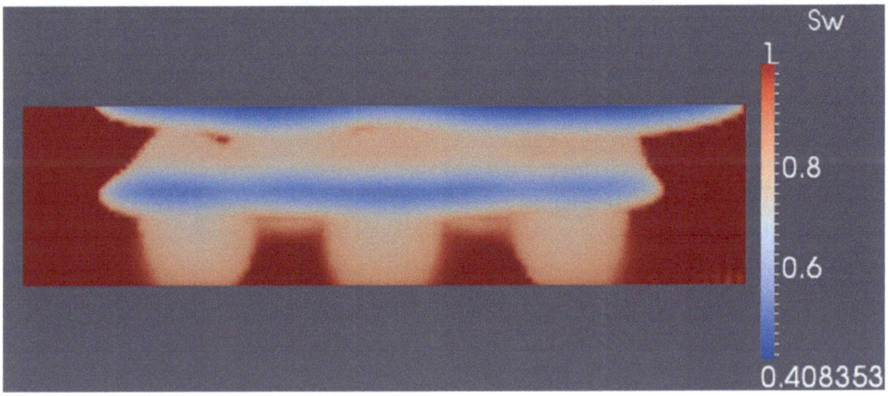

Fig. 5 Final saturation S_w with uniform $u_j = 3.6$

This shows that the early bang-bang strategy comes quite close to the optimal solution and is significantly superior to uniform injection. Accordingly, Figs. 3 and 4 look very similar, except for the red brine hole below the upper CO_2 plume in Fig. 3. Note that, in contrast to the uniform strategy in Fig. 5, the three lower "pillars" of nonwetting supercritical CO_2 in Figs. 3 and 4 are cut off from the upper portion of the supercritical plume, being immobilized below residual saturation $S_n^r = 0.1$.

We remark that in the optimal control scenario 27.91 % of the total amount of CO_2 are trapped at the final time. Among this trapped CO_2, the relative amount of residually trapped nonwetting CO_2 is 17.99 %.

The optimal injection rates in the upper and lower wells are shown in the respective Figs. 6 and 7. We recognize the following tendencies: The rates come rather close to "on–off" bang-bang strategies, while the upper wells switch back to maximal injection shortly before the end. This can be attributed to the fact that they

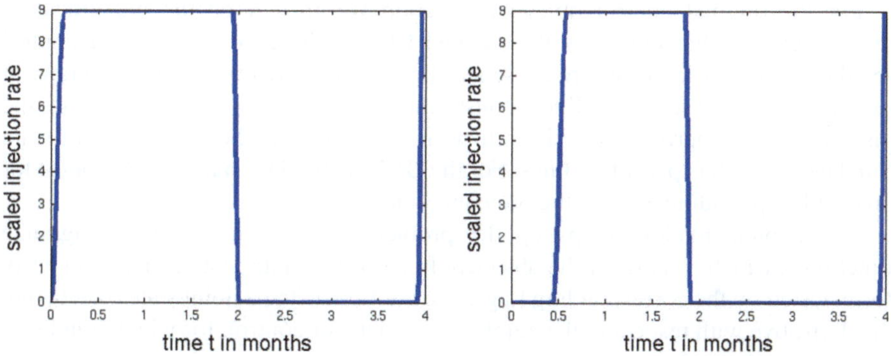

Fig. 6 Optimal rates in the upper two wells

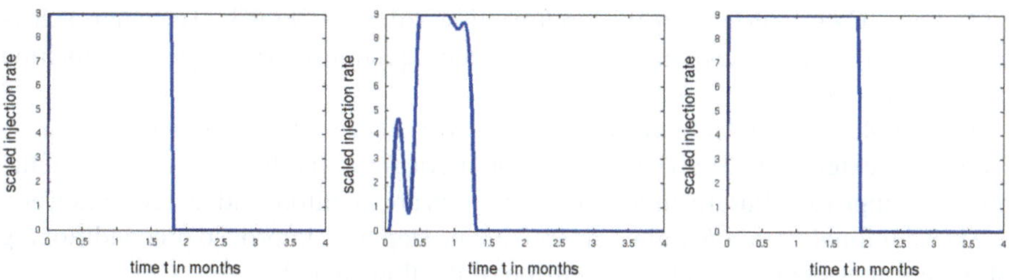

Fig. 7 Optimal rates in the lower three wells

are located in a layer of low permeability, so parts of the final portion of injected CO_2 can remain below residual saturation, due to a slower CO_2 plume evolution.

The mostly bang-bang like structure – i.e., either maximal or minimal injection in every well at all times ("on–off") – of the optimal strategies (with moderate deviation for the lower middle well) is supported by the recent article [44]: There it is shown that, given optimal control of EOR water-flooding with immiscible two-phase flow, such bang-bang solutions may occur if the control constraints on the injection rates are linear. Further optimization results for miscible two-phase flow in [37] agree with this observation.

7 Conclusions

In this paper, we investigated the optimal control of partially miscible multiphase flow in porous media. This setting is motivated by optimal CO_2 trapping in underground reservoirs. We introduced the multiphase flow model and described the BOX method that we use for discretization. The switching between the CO_2 mass fraction in the wetting phase and the saturation was formulated implicitly using a (for differentiability purposes smoothed) complementarity function.

For the optimal control objective function we considered the maximization of a functional that measures the trapped CO_2 at the final time plus a control regularization. The controls are chosen as the time-dependent CO_2 injection rates at multiple wells and subject to linear constraints. The derivative computation uses the discrete adjoint approach. The discretization was done based on the FEM software Sundance, which is part of Trilinos. For the BOX method we had to add support for dual cells, upwinding, etc., to the Sundance software package.

Our implementation wraps up the problem in the timestep-wise Lagrange functional, which consists of the objective functional and the state equation in weak form, tested by the corresponding Lagrange multiplier (the adjoint state). By taking the derivative with respect to the adjoint state, state, or control, for which Sundance provides automized support, operators for the (nonlinear) state time step equation, the (linear) adjoint time step equation and the (nonlinear) gradient of the reduced objective function are obtained. We linked the Sundance state and adjoint simulation code via a C++ interface to the optimization software IPOPT. The approach is quite flexible and makes changes in the state equation or the objective function comparably easy.

Our implementation is currently in 2D. We discussed which steps would be required to extend it to 3D. The presented numerical results show the viability of the proposed approach. Future directions of research are manifold and include extension to 3D, additional state constraints, adaptive time stepping, improved preconditioning of the linear systems, as well as extension of the flow model.

Acknowledgements This publication is based on work supported by Award No. UK-C0020, made by King Abdullah University of Science and Technology (KAUST). The work was conducted for the MAC-KAUST project K1 "Simulating CO_2 Sequestration" within the Munich Centre of Advanced Computing (MAC) at TUM. The authors gratefully acknowledge this support as well as the grant DFG INST 95/919-1 FUGG that provided partial funding of the compute cluster used for the computations. Moreover, the authors would like to thank Michael Bader for handling the paper and the three referees for their valuable comments.

References

1. Alexe, M., Sandu, A.: On the discrete adjoints of adaptive time stepping algorithms. J. Comput. Appl. Math. **233**(4), 1005–1020 (2009)
2. Asheim, H.: Maximization of water sweep efficiency by controlling production and injection rates. SPE paper 18365, presented at SPE European petroleum conference, London (1988)
3. Bastian, P.: Numerical computation of multiphase flows in porous media. Habilitation thesis, Christian-Albrechts-Universität Kiel (1999)
4. Ben Gharbia, I., Jaffre, J.: Gas phase appearance and disappearance as a problem with complementarity conditions. INRIA research report 7803 (2011)
5. Benk, J., Mehl, M., Ulbrich, M.: Sundance PDE solvers on Cartesian fixed grids in complex and variable geometries. In: Proceedings of ECCOMAS Thematic Conference on CFD and Optimization, Antalya (2011)

6. Benk, J., Bungartz, H.-J., Mehl, M., Ulbrich, M.: Immersed boundary methods for fluid-structure interaction and shape optimization within an FEM-based PDE toolbox. In: Bader, M., Bungartz, H.-J., Weinzierl, T. (eds.) Advanced Computing. Lecture Notes in CSE, vol. 93. Springer, New York (2013)
7. Bielinski, A.: Numerical simulation of CO_2 sequestration in geological formations. PhD thesis, Universität Stuttgart (2006)
8. Brandenburg, C., Lindemann, F., Ulbrich, M., Ulbrich, S.: A continuous adjoint approach to shape optimization for Navier Stokes flow. In: Kunisch, K., et al. (eds.) Optimal Control of Coupled Systems of Partial Differential Equations. International Series of Numerical Mathematics, vol. 158, pp. 35–56. Birkhäuser, Basel (2009)
9. Brandenburg, C., Lindemann, F., Ulbrich, M., Ulbrich, S.: Advanced numerical methods for PDE constrained optimization with application to optimal design in Navier Stokes flow. In: Engell, S., et al. (eds.) Constrained Optimization and Optimal Control for Partial Differential Equations, pp. 257–275. Birkhäuser, Basel (2011)
10. Brooks, R.H., Corey, A.T.: Hydraulic properties of porous media. Hydrology papers 3, Colorado State University, Fort Collins (1964)
11. Brouwer, D.R.: Dynamic water flood optimization with smart wells using optimal control theory. PhD thesis, Delft University of Technology (2004)
12. Brouwer, D.R., Jansen, J.D.: Dynamic optimization of water flooding with smart wells using optimal control theory. SPE J. **9**(4), 391–402 (2004)
13. Celia, M.A., Binning, P.: A mass-conservative numerical solution for two-phase flow in porous media with application to unsaturated flow. Water Resour. Res. **28**(10), 2819–2828 (1992)
14. Chen, Z., Huan, G., Ma, Y.: Computational Methods for Multiphase Flows in Porous Media. Computational Science and Engineering Series, vol. 2. SIAM, Philadelphia (2006)
15. Dolle, N., Brouwer, D.R., Jansen, J.D.: Dynamical optimization of water flooding with multiple injectors and producers using optimal control theory. In: Proceedings of 14th International Conference on Computational Methods in Water Resources, Delft (2002)
16. Enriquez, M.U.: The effects of coupling adaptive time-stepping and adjoint-state methods for optimal control problems. PhD thesis, Rice University, Houston (2010)
17. Fischer, A.: Solution of monotone complementarity problems with locally Lipschitzian functions. Math. Program. **76**(3B), 513–532 (1997)
18. Gao, G., Reynolds, A.C.: An improved implementation of the LBFGS algorithm for automatic history matching. SPE J. **11**(1), 5–17 (2006)
19. Gockenbach, M.S., Symes, W.W.: Adaptive simulation, the adjoint state method, and optimization. In: Biegler, L.T., et al. (eds.) Large Scale PDE-Constrained Optimization, pp. 281–297. Springer, Berlin/New York (2003)
20. Helmig, R.: Multiphase Flow and Transport Processes in the Subsurface. Springer, Berlin (1997)
21. Helmig, R., Niessner, J., Class, H.: Recent advances in finite element methods for multiphase flow processes in porous media. Int. J. Comput. Fluid Dyn. **20**(3), 245–252 (2006)
22. Heroux, M.A., Willenbring, J.M., Heaphy, R.: Trilinos developers guide. Sandia National Laboratories, SAND 2003-1898 (2003)
23. Hinze, M., Sternberg, J.: A-revolve: an adaptive memory-reduced procedure for calculating adjoints; with an application to computing adjoints of the instationary Navier-Stokes system. Optim. Methods Softw. **20**(6), 645–663 (2005)
24. Hinze, M., Pinnau, R., Ulbrich, M., Ulbrich, S.: Optimization with PDE Constraints. Mathematical Modelling: Theory and Applications, vol. 23. Springer, New York (2008)
25. Huber, R., Helmig, R.: Multiphase flow in heterogeneous porous media: a classical finite element method versus an implicit pressure–explicit saturation-based mixed finite element–finite volume approach. Int. J. Numer. Methods Fluids **29**(8), 899–920 (1999)
26. Jansen, J.D.: Adjoint-based optimization of multi-phase flow through porous media – a review. Comput. Fluids **46**(1), 40–51 (2011)
27. Lauser, A., Hager, C., Helmig, R., Wohlmuth, B.: A new approach for phase transitions in miscible multi-phase flow in porous media. Adv. Water Resour. **34**(8), 957–966 (2011)

28. Li, S., Petzold, L.: Adjoint sensitivity analysis for time-dependent partial differential equations with adaptive mesh refinement. J. Comput. Phys. **198**(1), 310–325 (2004)
29. Lindemann, F.: Theoretical and Numerical Aspects of Shape Optimization with Navier-Stokes Flows. PhD thesis, Technische Universität München (2012)
30. Long, K.R.: Sundance rapid prototyping tool for parallel PDE optimization. In: Biegler, L.T., et al. (eds.) Large Scale PDE-Constrained Optimization, pp. 331–342. Springer, Berlin/New York (2003)
31. Long, K.R., Boggs, P.T., van Bloemen Waanders, B.G.: Sundance: high-level software for PDE-comstrained optimization. Sci. Program. **20**(3), 293–310 (2012)
32. Mehos, G.J., Ramirez, W.F.: Use of optimal control theory to optimize carbon dioxide miscible-flooding enhanced oil recovery. J. Pet. Sci. Eng. **2**(4), 247–260 (1989)
33. Neitzel, I., Tröltzsch, F.: On convergence of regularization methods for nonlinear parabolic optimal control problems with control and state constraints. Control Cybern. **37**(4), 1013–1043 (2008)
34. Oliver, D.S., Reynolds, A.C., Liu, N.: Inverse Theory for Petroleum Reservoir Characterization and History Matching. Cambridge University Press, Cambridge (2008)
35. Peszynska, M., Jenkins, E.W., Wheeler, M.F.: Boundary conditions for fully implicit two-phase flow models. In: Feng, X., Schulze, T.P. (eds.) Recent Advances in Numerical Methods for Partial Differential Equations and Applications. Contemporary Mathematics Series, vol. 306, pp. 85–106. AMS, Providence (2002)
36. Sarma, P., Aziz, K., Durlofsky, L.J.: Implementation of adjoint solution for optimal control of smart wells. SPE paper 92864, presented at SPE reservoir simulation symposium, Houston (2005)
37. Simon, M., Ulbrich, M.: Adjoint based optimal control of partially miscible two-phase flow in porous media with applications to CO_2 sequestration in underground reservoirs. Preprint, Technische Universität München (2013)
38. Sudaryanto, B., Yortsos, Y.C.: Optimization of fluid front dynamics in porous media using rate control. Phys. Fluids **12**(7), 1656–1670 (2000)
39. Sudaryanto, B., Yortsos, Y.C.: Optimization of displacements in porous media using rate control. SPE paper 71509, presented at SPE annual technical conference and exhibition, New Orleans (2001)
40. Ulbrich, M.: Semismooth Newton Methods for Variational Inequalities and Constrained Optimization Problems in Function Spaces. MOS-SIAM Series on Optimization, vol. 11. SIAM, Philadelphia (2011)
41. Virnovski, G.A.: Water flooding strategy design using optimal control theory. In: Proceedings of 6th European Symposium on IOR, Stavanger (1991)
42. Wächter, A., Biegler, L.T.: On the implementation of a primal-dual interior point filter line search algorithm for large-scale nonlinear programming. Math. Program. **106**, 25–57 (2006)
43. Zakirov, I.S., Aanonsen, S.I., Zakirov, E.S., Palatnik, B.M.: Optimization of reservoir performance by automatic allocation of well rates. In: Proceedings of ECMOR V: European Conference on Mathematics of Oil Recovery, Leoben (1996)
44. Zandvliet, M.J., Bosgra, O.H., Jansen, J.D., van den Hof, P.M.J., Kraaijewanger, J.F.B.M.: Bang-bang control and singular arcs in reservoir flooding. J. Pet. Sci. Eng. **58**(1), 186–200 (2007)
45. Zhang, Q.: A finite difference–streamline diffusion (FDSD) method and its error estimates for two-phase incompressible miscible flow in porous media. Acta Math. Appl. Sin. **26**(2), 318–327 (2003)
46. Zhu, C., Byrd, R.H., Nocedal, J.: L-BFGS-B: Algorithm 778: L-BFGS-B, FORTRAN routines for large scale bound constrained optimization. ACM Trans. Math. Softw. **23**(4), 550–560 (1997)

A Newton-CG Method for Full-Waveform Inversion in a Coupled Solid-Fluid System

Christian Boehm and Michael Ulbrich

Abstract We present a Newton-CG method for full-waveform seismic inversion. Our method comprises the adjoint-based computation of the gradient and Hessian-vector products of the reduced problem and a preconditioned conjugate gradient method to solve the Newton system in matrix-free fashion. A trust-region globalization strategy and a multi-frequency inversion approach are applied. The governing equations are given by a coupled system of the acoustic and the elastic wave equation for the numerical simulation of wave propagation in solid and fluid media. We show numerical results for the application of our method to marine geophysical exploration.

Keywords Seismic tomography • Full-waveform inversion • Elastic-acoustic coupling • Newton-CG • Trust-region globalization • Goal-oriented adaptivity

1 Introduction

Earthquakes excite seismic waves that propagate through the Earth and can be recorded as seismograms at remote receiver locations. Seismic tomography means to infer the Earth's structure based on these observations. An accurate knowledge of the Earth's interior does not only enhance scientific progress in explaining the geodynamics and subsurface processes but can also help to improve reliable

C. Boehm (✉)
Department of Mathematics, Technische Universität München, Boltzmannstr. 3, 85748 Garching b. München, Germany
e-mail: boehm@ma.tum.de

M. Ulbrich
Chair of Mathematical Optimization, Department of Mathematics, Technische Universität München, Boltzmannstr. 3, 85748 Garching b. München, Germany
e-mail: mulbrich@ma.tum.de

M. Bader et al. (eds.), *Advanced Computing*, Lecture Notes in Computational Science and Engineering 93, DOI 10.1007/978-3-642-38762-3_5,

Tsunami warning systems and support the search for natural resources. In this article we focus on the application of marine geophysical exploration, where seismic waves are emitted by a research vessel that is equipped with an air gun and cruises on the sea. This requires to model a medium consisting of a solid and a fluid layer as well as the interaction at the interface. The governing equations are given by a coupled system of the elastic wave equation in the solid domain and the acoustic wave equation in the fluid domain.

Seismic tomography can be stated as an optimization problem with PDE constraints. Here, the spatially heterogeneous, unknown material parameters enter the PDE as coefficients. We refer to [34] for a general overview on seismic tomography. With the availability of high-performance computing clusters, 3D full-waveform inversion has become possible and iterative inversion methods have been applied to datasets on both, regional and continental scale [12, 14, 15, 37, 38]. Alternative approaches work in the frequency domain and involve the Helmholtz equation [31, 41]. The majority of the presented inversion algorithms relies on gradient-based methods but extensions that incorporate second-order information have been made [8, 12, 32].

Here, we present a Newton-type method for full-waveform inversion. We apply a trust-region globalization and iteratively solve the resulting subproblems by the Steihaug-CG method [36]. The gradient and Hessian-vector products of the reduced problem are efficiently computed using adjoint-based techniques [21]. A smooth cutoff function ensures that the parameters remain within reasonable bounds without explicitly imposing additional constraints. We invert sequentially for increasing source frequencies and adaptively refine the parameter grid using goal-oriented error estimates [4]. The coupled system is spatially discretized by a high-order continuous Galerkin method and solved with an explicit Newmark time-stepping scheme [28, 30]. The parallel implementation works matrix-free and utilizes MPI communication to tackle large-scale seismic inverse problems.

This article is organized as follows. In Sect. 2 we describe the governing equations and formulate the seismic inverse problem. In Sect. 3 we present our optimization method and conclude with some remarks on the implementation and a numerical example in Sect. 4.

2 The Seismic Inverse Problem

2.1 Wave Propagation at a Solid-Fluid Interface

We consider a domain that consists of a solid and a fluid layer. We denote the solid and fluid regions by Ω_S and Ω_F and set $\Omega = \Omega_S \cup \Omega_F \subset \mathbb{R}^d$ with $d = 2, 3$. We assume that Ω_S and Ω_F are bounded domains with sufficiently regular boundaries and interface Γ_{int}. The remaining parts of the boundaries are denoted by $\Gamma_F = \partial\Omega_F \setminus \Gamma_{\text{int}}$ and $\Gamma_S = \partial\Omega_S \setminus \Gamma_{\text{int}}$. The time interval is denoted by $I := (0, T)$ with $T > 0$.

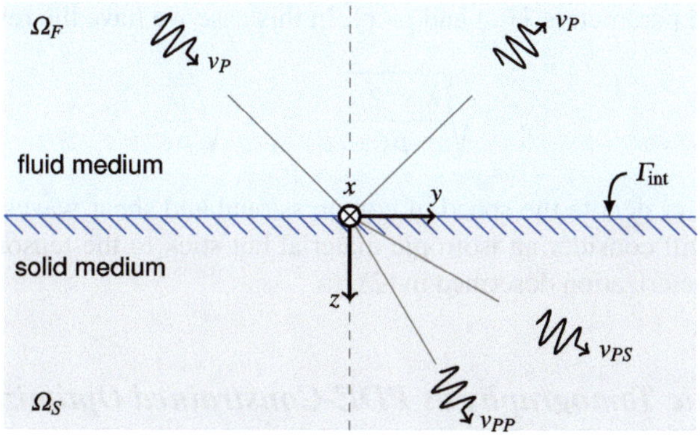

Fig. 1 Sketch of a solid-fluid interface. Parts of the incoming waves are reflected to the acoustic medium while other parts are transmitted to the elastic medium

The propagation of waves in the solid medium is governed by the elastic wave equation and, respectively, by the acoustic wave equation in the fluid domain. At the interface, continuity of traction and continuity of the normal displacement have to be ensured, cf. [9,28]. Figure 1 shows the geometry of the domain with both layers.

In the fluid domain, we consider an inviscid fluid medium with a homogeneous density $\rho_F > 0$. χ denotes the displacement potential and $c \in \mathbb{R}$ is the speed of compressional waves. In the solid medium, we assume a heterogeneous and positive density $\rho_S(x)$ and a linear elastic rheology. Let u denote the displacement field, $\varepsilon(u) = \frac{1}{2}\left(\nabla u + \nabla u^T\right)$ the strain tensor of u and $\Psi = (\Psi_{ijkl})$ is a fourth order elastic tensor. To cover the general case, we place seismic source functions f_F and f_S into both media. The coupled system is then given by:

$$\left\{\begin{array}{rll}
\rho_F \chi_{tt}(x,t) - \rho_F c^2 \Delta\chi(x,t) &= f_F(x,t) & (x,t) \in \Omega_F \times I,\\
\chi(x,0) = 0, \quad \chi_t(x,0) &= 0 & x \in \Omega_F,\\
\chi(x,t) &= 0 & (x,t) \in \Gamma_F \times I,\\
&&\\
\rho_S(x) u_{tt}(x,t) - \nabla\cdot(\Psi(x) : \varepsilon(u)(x,t)) &= f_S(x,t) & (x,t) \in \Omega_S \times I,\\
u(x,0) = 0, \quad u_t(x,0) &= 0 & x \in \Omega_S,\\
(\Psi(x) : \varepsilon(u)(x,t)) \cdot \mathbf{n}_S &= 0 & (x,t) \in \Gamma_S \times I,\\
&&\\
(\Psi(x) : \varepsilon(u)(x,t)) \cdot \mathbf{n}_S &= -\rho_F \chi_{tt}(x,t)\, \mathbf{n}_F & (x,t) \in \Gamma_{\text{int}} \times I,\\
-\mathbf{n}_F \cdot \nabla\chi(x,t) &= \mathbf{n}_S \cdot u(x,t) & (x,t) \in \Gamma_{\text{int}} \times I.
\end{array}\right. \tag{1}$$

Remark 1. The tensor Ψ of elastic moduli has the symmetry properties $\Psi_{ijkl} = \Psi_{jikl} = \Psi_{klij}$ and has to be uniformly coercive. In this general form we can handle anisotropic material. However, in a perfectly elastic, isotropic medium the tensor simplifies to

$$\Psi_{ijkl} = \lambda\delta_{ij}\delta_{kl} + \mu(\delta_{ik}\delta_{jl} + \delta_{il}\delta_{kj}) \tag{2}$$

with the Lamé parameters $\lambda(x)$ and $\mu(x)$. In this case we have the relation

$$v_p = \sqrt{\frac{\lambda + 2\mu}{\rho_S}}, \quad v_s = \sqrt{\frac{\mu}{\rho_S}}, \tag{3}$$

where v_p and v_s denote the speed of compressional and shear waves [29]. In what follows, we will consider an isotropic material but stick to the tensor notation and use the parameterization described in (2).

2.2 *Seismic Tomography as PDE-Constrained Optimization Problem*

Depending on the parameterization of the governing equations, the unknown material parameters can be the Lamé parameters, the velocity of compressional and/or shear waves or further elasticity parameters. In either case, the unknown parameter field is heterogeneous in space and does not depend on the time. Due to the interdependencies of ρ_S and Ψ, we keep the density fixed and invert for Ψ only. Since the material structure can be quite irregular in general, we want to work with mild assumptions on the function space and prefer to use $L^\infty(\Omega)^{d^4}$. On the other hand, we require a higher regularity for the regularization term to treat the ill-posedness of the problem. In order to overcome this tradeoff, we split the material properties into a reference model $\overline{\Psi} \in L^\infty(\Omega_S)^{d^4}$ that is based on a-priori knowledge and the material variable m that parameterizes smooth variations from the reference model. Let n denote the number of components in the unknown parameter field, $M \subset\subset L^\infty(\Omega_S)^n$ a Hilbert space and $\Phi : \mathbb{R}^n \to \mathbb{R}^{d^4}$ a linear function. The complete parameterization is then given by

$$\Psi(m) = \overline{\Psi} + \Phi(m). \tag{4}$$

Remark 2. In all applications that we are concerned with, the existence of a suitable reference model based on a-priori knowledge is guaranteed. Often, those reference models only vary in depth and assume a homogeneous parameter field in the horizontal plane. For instance, the most popular choice for global seismic tomography is the Preliminary Reference Earth Model (PREM) [11].

In the next step, we turn to the weak formulation of (1). To shorten the notation, we define the following forms:

$$\begin{aligned}
a_S(m)(v,w) &= (\Psi(m) : \varepsilon(v), \varepsilon(w))_{L^2(\Omega_S)^{d\times d}} && \forall\, v,w \in V_S,\ \forall\, m \in M,\\
a_F(v,w) &= \rho_F c^2\, (\nabla v, \nabla w)_{L^2(\Omega_F)^d} && \forall\, v,w \in V_F,\\
a_{\text{int}}(v,w) &= \rho_F \int_{\Gamma_{\text{int}}} w\, v \cdot \mathbf{n}_s\, dS && \forall\, v \in L^2(\Omega_S)^d,\ \forall\, w \in L^2(\Omega_F).
\end{aligned}$$

Here, $V_S = H^1(\Omega_S)^d$ and $V_F = H_0^1(\Gamma_F, \Omega_F)$ denotes the space of all H^1-functions that vanish on Γ_F. The normal vector $\mathbf{n}_S$ in the definition of a_{int} is pointing outwards of the solid and into the fluid domain. The variational form of (1) is given by: $\forall\, v \in V_F, \forall\, w \in V_S$ and a.a. $t \in I$:

$$\begin{aligned} \rho_F \langle \chi_{tt}(t), v\rangle_{V_F^*,V_F} + a_F(\chi(t), v) + c^2\, a_{\text{int}}(u(t), v) &= \langle f_F(t), v\rangle_{V_F^*,V_F}, \\ \langle \rho_S\, u_{tt}(t), w\rangle_{V_S^*,V_S} + a_S(m)(u(t), w) - a_{\text{int}}(w, \chi_{tt}(t)) &= \langle f_S(t), w\rangle_{V_S^*,V_S}. \end{aligned} \tag{5}$$

The dual spaces of V_S and V_F are denoted by V_S^* and V_F^*, respectively. In [5] we discuss existence and uniqueness of solutions in appropriate function spaces. This analysis, however, is beyond the scope of this article, so we just state the following concepts for fixed $m \in M$:

- For the acoustic wave equation with homogeneous Dirichlet boundary conditions and a source $f_F \in L^2(I, V_F^*)$ there exists a unique very weak solution $\chi \in X \overset{\text{def}}{=} C(\bar{I}; L^2(\Omega_F)) \cap C^1(\bar{I}; V_F^*)$.
- For the elastic wave equation with homogeneous Neumann boundary conditions and a source $f_S \in L^2(I, V_S^*)$ there exists a unique very weak solution $u \in U \overset{\text{def}}{=} C(\bar{I}; L^2(\Omega_S)^d) \cap C^1(\bar{I}; V_S^*)$.

Differentiability of u with respect to m can be established by exploiting higher time regularity of the seismic source function.

Remark 3. Note that a higher time regularity is a valid assumption for the seismic source function as wavelets are commonly used. Since we can admit V_F^* or V_S^* in space, Dirac measures require only a slight smoothing.

We set $Y = U \times X$ and let $y = (u, \chi)$ denote the state that consists of the displacement field in the solid domain and the displacement potential in the fluid domain. With $f = (f_F, f_S)^T$ the weak form of the coupled system (1) can be written as:

$$E(y, m) = f \quad :\Leftrightarrow (y, m) \text{satisfies (5) a.e. in } I. \tag{6}$$

Additionally, the initial conditions $y(0) = 0$ and $y_t(0) = 0$ have to be satisfied.

In the application of marine geophysical exploration, data is taken from several seismic events, which means that we have to solve a wave equation with a different right hand side for every event independently. We denote the number of events by n_s and the corresponding state variables by y_i, $i = 1, \ldots, n_s$. For every i, we are given a seismic source $f_i = (f_F)_i$ with support only in the fluid domain. Furthermore, we assume to have measurements y_i^δ of the displacement field in form of seismograms on $\Omega_i \times I$ with $\Omega_i \subset \Omega_S$. Typically, only sparse observations of the displacement field are available which contributes to the ill-posedness of the problem and necessitates proper regularization. The seismic inverse problem is given as follows

$$\min_{\mathbf{y}\in\mathbf{Y},\, m\in M} J(\mathbf{y},m) \quad \text{s.t. } E(y_i,m) = \begin{pmatrix} f_i \\ 0 \end{pmatrix} \quad 1 \le i \le n_s, \quad \begin{pmatrix} \mathbf{y}(0) \\ \mathbf{y}_t(0) \end{pmatrix} = 0. \tag{7}$$

Here, $\mathbf{y} = (y_i)_{1\le i\le n_s}$ denotes a vector of states for different seismic events. Note that the state y_i only enters into the i-th component of E, while the parameters m are the same for all components. We consider cost functions $J : \mathbf{Y} \times M \to \mathbb{R}$ of the form:

$$J(\mathbf{y},m) = \sum_{i=1}^{n_s} J_{\text{fit}}(y_i, y_i^\delta) + \alpha J_{\text{reg}}(m), \tag{8}$$

with a misfit term J_{fit}, a regularization term J_{reg} and the regularization parameter $\alpha > 0$. In what follows, we will always assume that J_{fit} is twice continuously F-differentiable with respect to y_i, and J_{reg} is twice continuously F-differentiable with respect to m, convex and lower semicontinuous. A possible choice for J would be, for instance,

$$J_{\text{fit}}(y_i, y_i^\delta) = \frac{1}{2}\|y_i - y_i^\delta\|^2_{L^2(\Omega_i\times I)}, \quad J_{\text{reg}}(m) = \frac{1}{2}\|m\|_M^2. \tag{9}$$

Furthermore, we introduce the reduced problem

$$\min_{m\in M} j(m) \stackrel{\text{def}}{=} J(\mathbf{y}(m),m), \tag{10}$$

where $\mathbf{y}(m)$ denotes the solution of the state equation for given $m \in M$.

A significant extension to the problem formulation (7) can be made if additional box constraints on the material parameters are imposed. This can also be found in [5].

Remark 4. In this paper, we consider the inverse problem to determine the spatially heterogeneous material parameters and assume that the location and time evolution of the seismic source f is known. However, our optimization method that is outlined in Sect. 3 could conceptually be extended to invert for the seismic source as well (either simultaneously or in an alternating manner).

2.3 Adjoint Equation

As pointed out above, the state variables for different seismic events completely decouple in (7). Hence, we restrict the analysis of the adjoint equation to a single seismic event and drop the index i. Derivatives of the multi-source problem can be computed by adding up the individual contributions. In order to derive the adjoint equation, we introduce the Lagrange function $L(y,m,z) : Y \times M \times Y \to \mathbb{R}$ with the adjoint variable $z = (z^1, z^2)$:

$$
\begin{aligned}
L(y,m,z) \stackrel{\text{def}}{=} & J(y,m) + \int_0^T \langle \rho_S\, u_{tt}(t), z^1(t)\rangle_{V_S^*,V_S}\, dt + \int_0^T a_S(m)(u(t), z^1(t))\, dt \\
& + \int_0^T \rho_F \langle \chi_{tt}(t), z^2(t)\rangle_{V_F^*,V_F}\, dt + \int_0^T a_F(\chi(t), z^2(t))\, dt \\
& - \int_0^T a_{\text{int}}(z^1, \chi_{tt}(t))\, dt + \int_0^T c^2\, a_{\text{int}}(u(t), z^2(t))\, dt \\
& - \int_0^T \langle f_S(t), z^1(t)\rangle_{V_S^*,V_S}\, dt - \int_0^T \langle f_F(t), z^2(t)\rangle_{V_F^*,V_F}\, dt \\
& -(\rho_S\, u(0), z_t^1(0))_{L^2(\Omega_S)^d} + (\rho_S\, u_t(0), z^1(0))_{L^2(\Omega_S)^d} \\
& -\rho_F(\chi(0), z_t^2(0))_{L^2(\Omega_F)} + \rho_F(\chi_t(0), z^2(0))_{L^2(\Omega_F)}.
\end{aligned}
\tag{11}
$$

The adjoint equation is given by

$$L_y(y,m,z) = 0. \tag{12}$$

By considering the variational form of (12) and carefully integrating by parts with respect to time, we obtain the adjoint equation for given m and $y(m)$ as

$$E^{\text{ad}}(z,m) = -J_y(y(m),m), \quad z(T) = 0, \quad z_t(T) = 0. \tag{13}$$

With the assumption of sufficient regularity, the adjoint equation E^{ad} in strong form can be interpreted as a coupled system like (1) backwards in time with final time instead of initial conditions, interchanged interface conditions and a different right-hand-side. Further details of deriving the continuous adjoint can be found in [5].

In the next step, we turn to the adjoint-based representation of first and second derivatives. Using the parameterization defined in (4) we obtain

$$\Psi'(m) = \Phi(.) \in \mathscr{L}\left(M, L^\infty(\Omega_S)^{d^4}\right) \quad \forall\, m \in M. \tag{14}$$

We introduce the form $D : Y \times Y \to \mathscr{L}(M,\mathbb{R})$ defined by

$$D(v,w)(m) = \int_0^T \int_{\Omega_S} \left(\varepsilon(v^1)(x,t) \otimes \varepsilon(w^1)(x,t)\right) :: (\Phi(m)(x))\; dx\, dt \tag{15}$$

for all $m \in M$. Here, $v = (v^1, v^2), w = (w^1, w^2) \in U \times X$ and we use the notations $(a \otimes b)_{ijkl} = a_{ij} b_{kl}$ and $A :: B = \sum_{ijkl} A_{ijkl} B_{ijkl}$. For given $m \in M$ we denote the corresponding state by $y(m)$ and the adjoint state by $z(m)$. The first derivatives can then be expressed as

$$j'(m) = \alpha J'_{\text{reg}}(m) + D(y(m), z(m)). \tag{16}$$

Following the derivation in [21], the second derivative of the reduced cost functional can be represented in a similar fashion. This requires the second derivatives of the Lagrange function with respect to the state and the parameters. Since the state equation is linear in both, y and m, a couple of terms vanish. While the operator $j''(m)$ would be prohibitively expensive to compute, operator-vector products $j''(m)s$ for a given perturbation $s \in M$ can be computed at the cost of two additional simulations by performing the following steps:

1. Compute a perturbed forward wavefield $\delta_s y$ by solving

$$E(\delta_s y, m) = -E_m(y(m), m)s, \quad \delta_s y(0) = 0, \quad \delta_s y_t(0) = 0. \tag{17}$$

2. Compute a perturbed adjoint wavefield $\delta_s z$ by solving

$$E^{\mathrm{ad}}(\delta_s z, m) = -J_{yy}\delta_s y - E_m(z(m), m)s, \quad \delta_s z(T) = 0, \quad \delta_s z_t(T) = 0. \tag{18}$$

Here, E_m denotes the linearization of E with respect to m. $j''(m)s$ is then given by:

$$j''(m)s = \alpha J''_{\mathrm{reg}}(m)s + D(\delta_s y, z(m)) + D(y(m), \delta_s z). \tag{19}$$

With the adjoint-based representation of the first derivatives and operator-vector products representing the second derivatives applied to a search direction, we have everything at hand to apply a Newton-type optimization method.

3 Optimization Method

3.1 *Trust-Region Newton-CG*

In order to solve the seismic inverse problem (10), the algorithm iteratively computes approximate solutions to the trust-region subproblem

$$\begin{aligned} \min_{s \in M} \quad & q_i(s) \stackrel{\mathrm{def}}{=} j(m^i) + \langle j'(m^i), s\rangle_{M^*,M} + \tfrac{1}{2}\langle j''(m^i)s, s\rangle_{M^*,M} \\ & \text{s.t. } \|s\|_M \leq \Delta_i. \end{aligned} \tag{20}$$

Here m^i denotes the current iterate and Δ_i the trust-region radius in iteration i. q_i is a quadratic model function that approximates $j(m^i + s)$. The first derivatives $j'(m^i)$ and operator-vector products $j''(m^i)s$ are computed using adjoint-based techniques as outlined in the previous section. Note that we use the exact second derivatives in (20), however, approximations of the Hessian, e.g. by a Quasi-Newton method, would also be possible. We compute an approximate solution to (20) by the Steihaug conjugate gradient method [36]. Here, the inner product induced by the norm of M is used as preconditioner. The CG iterations are terminated early, if negative curvature is encountered or the trust region radius is exceeded by the current iterate. Additionally, we solve the Newton system inexactly and stop with a relative tolerance of 0.01. A crucial ingredient of the trust-region method is the ratio of actual and predicted reduction,

$$\frac{j(m^i) - j(m^i + s^i)}{q_i(0) - q_i(s^i)}, \tag{21}$$

based on which is determined, whether the step is accepted and how the trust-region radius is updated. For further details, we refer to [5], see also [40] for trust-region methods in function space and [10] for a comprehensive study.

The problem formulation (7) does not include any constraints on the material parameters. However, from a physical, as well as from a theoretical point of view, there exist bounds on m, for instance, to ensure nonnegative wave velocities or coercivity of the elastic tensor. In [5] we extend the problem formulation by imposing additional pointwise box constraints on the material parameters. The trust-region algorithm presented above is then applied to a series of problems with an additional penalty term in the objective function that is weighted with a monotonically increasing penalty parameter. For the simplified unconstrained case that is discussed in this article, we apply a smooth cutoff function to ensure that the parameters remain within a certain range. Let $m^{l-}, m^l, m^u, m^{u+} \in \mathbb{R}^n$, $m^{l-} < m^l < m^u < m^{u+}$, denote pointwise bounds on the material parameters. We then choose a smooth function $\varphi : M \rightarrow M$ and replace m by $\varphi(m)$ in the weak form (5). Hereby, φ has to satisfy the following properties: φ is monotonically increasing, $\varphi \equiv \text{id}$ on $[m^l, m^u]$ and $\varphi(M) = [\varphi(m^{l-}), \varphi(m^{u+})]$. The bounds should be chosen such that $[\varphi(m^{l-}), \varphi(m^{u+})]$ covers the domain of all physically feasible models and the optimal parameter model should be within $[m^l, m^u]$. Thus, it is guaranteed that only reasonable values for m enter into the weak form. Moreover, the cutoff ensures that parameter models that exceed the bounds yield the same misfit while the regularization term J_{reg} penalizes large deviations. Hence, the optimization algorithm will eventually favor parameter models within the bounds. Figure 2 shows an example of this cutoff function φ.

3.2 *Discretization*

We follow a discretize-then-optimize strategy. This involves the temporal and spatial discretization of the state and adjoint equation as well as the spatial discretization of the parameters. We use different meshes for the state and the parameter space. This is motivated by the fact that the information on the material properties is limited, thus, a coarser mesh in the parameter space prevents an over-parameterization. This regularization-by-discretization strategy is enhanced with an adaptive grid refinement of the parameter space in combination with a multi-frequency inversion approach which is explained in the next section.

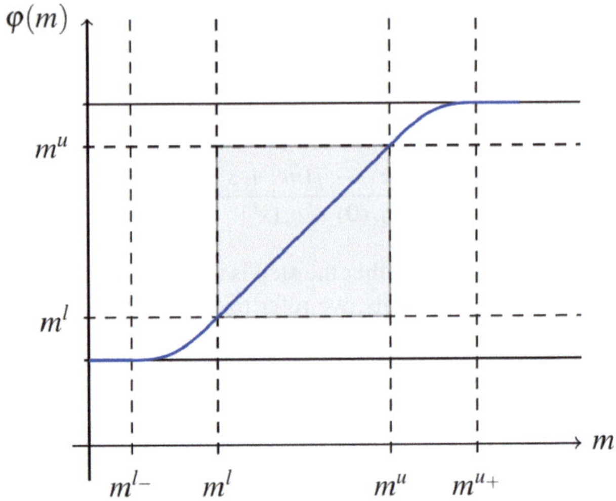

Fig. 2 Smooth cutoff function to ensure that the parameters remain within physically reasonable bounds. A fourth order polynomial ensures the smooth transition in $[m^{l-}, m^l]$ and $[m^u, m^{u+}]$

We apply a continuous high-order finite element method for the spatial discretization of the state space and an explicit Newmark time-stepping. We use fourth-order Lagrange polynomials and approximate the integrals by the Legendre-Gauss-Lobatto quadrature rule which results in a diagonal mass matrix. This approach is commonly used in seismic applications, cf. [13, 39, 42]. Moreover, by following an update scheme for the Newmark time-stepping as outlined in [28], we can carry out the time-stepping fully explicit.

The seismic inverse problem is computationally very expensive. The costs are vastly dominated by solving the discrete version of the coupled system (1) for several right-hand-sides. $2n_s$ simulations (one forward, one adjoint per seismic event) are required to evaluate the cost function and to compute the first derivatives for a given parameter model m. Every inner CG iteration requires additional $2n_s$ simulations to compute the second derivatives applied to a search direction. In order to limit the computational costs of a Newton-step in comparison to a limited-memory BFGS approximation, we restrict the number of CG steps per outer iteration to 20, in addition to the stopping criterion based on the relative reduction of the residual.

In order to further reduce the number of simulations per iteration, randomized data reduction techniques are an interesting field of research, cf. [1, 24]. Here, the independent seismic events are replaced by simultaneous "super-shots".

3.3 Multi-frequency Approach

A reconstruction of the material properties with a high resolution requires high frequency information in the observed data. However, the high-frequency data is more prone to errors induced by noisy measurements. Moreover, since only sparse observations are available, there usually exist several models that explain the data equally well. This necessitates the regularization term in the objective function. Additionally, we follow a regularization-by-discretization strategy and combine a multi-frequency inversion approach with an adaptive grid refinement based on goal-oriented error estimates. The multi-frequency approach (sometimes also called multi-scale) means to sequentially invert for increasing source frequencies [7, 16], see also [12]. A bandpass filter can be used to down-sample the observed measurements to lower frequencies. The adaptively refined grid allows to reduce the number of optimization variables without a loss of resolution in the reconstruction. The algorithm works as follows:

0. Choose an initial parameter mesh.

For a sequence of increasing source frequencies $\omega_1 \leq \omega_2 \leq \dots$

1. Choose the state mesh based on the dominant frequency ω_i of the seismic source and the wave velocities.
2. Solve the discretized problem and obtain a stationary point.
3. Adaptively refine the parameter mesh using goal-oriented error estimators.

The paradigm for goal-oriented a-posteriori error estimation and dual weighted residuals was established in [4]. We briefly outline the key idea. Let $m^* \in M$ be a stationary point of the Lagrange function L for the continuous problem and $m_h \in M_h \subset M$ a stationary point of L on the discretized subspace M_h. Then the error in the cost function can be represented by

$$j(m^*) - j(m_h) = \frac{1}{2}\langle L_m(y(m_h), m_h, z(m_h)), m^* - v_h\rangle_{M^*,M} + R, \tag{22}$$

with a cubic remainder term R and an arbitrary $v_h \in M_h$. In order to compute error estimates, the derivative of the Lagrange function with respect to m has to be computed which requires a state and an adjoint equation solve. A remaining difficulty is to estimate the difference $m^* - v_h$, since the continuous solution m^* is unknown. Here, we use higher-order local interpolation in a post-processing step. We refer to [3, 25, 27] for goal-oriented error estimation in general and to [5] for a detailed description of its application in the context of seismic tomography.

4 Numerical Example

4.1 Implementation

Efficient inversion methods rely on a scalable code for the simulation of the elastic-acoustic wave equation. We have implemented the wave propagation code in C++, following a similar approach like SPECFEM3D [30]. Parallelization is

Table 1 Strong parallel scaling statistic. The table indicates the speedup compared to the run on four processors for a forward simulation with one seismic source. n_p denotes the number of processors. The problem size is given by 128,000 elements in space and 1,000 time steps

n_p	4	8	16	32	64	128
Speedup	1.0	2.06	3.41	4.98	8.76	10.42

carried out in two stages. Trivially, different seismic events can be simulated in parallel and communication is only required during a post-processing step to add up the individual contributions to the cost functional and its derivatives. Hence, including several seismic events scales almost perfectly. Moreover, our parallel implementation allows to solve a single event on multiple cores using a spatial partitioning of the computational domain and communication with MPI. The algorithm works matrix-free and does not require to solve a linear system during the simulation. We use the Epetra package of Trilinos [20] to handle the distributed data structures.

Table 1 shows the strong scaling statistics for a forward simulation of a single seismic event. In near future, we plan to increase the parallel efficiency by exploiting non-blocking MPI communication.

The adaptive refinement of the parameter mesh is implemented using the deal.ii library [2]. The parameter mesh is then interpolated onto the state mesh before every simulation.

The computation of the first derivatives requires the forward displacement field at all time-steps. Moreover, the two wave equations (17) and (18), that have to be solved for the Hessian-vector product, additionally require the adjoint state at all time-steps. In the current setup, we store both wavefields, however, if memory poses a bottleneck, checkpointing techniques [18] could be applied. Note that we do not have to store the perturbed wavefields $\delta_s y$ and $\delta_s z$, because the contributions to the derivatives can be computed on the fly.

4.2 Acoustic-Elastic Data Set

This example shows the application of our method to a synthetic data set in marine geophysical exploration. The geometry and problem setup (Fig. 3) is inspired by the Valhall oil field in the North Sea. Prior work on this field can be found in [6,33,35]. The geometry is given by a rectangular domain of $8 \times 8 \times 4$ km.

There is a thin water layer (400 m) on top of the solid domain. We use 36 seismic sources that are triggered simultaneously in the fluid region at 200 m water depth. The source time function for all sources is a Ricker wavelet with dominant frequency of 2.5 Hz. There are 441 seismic receivers buried into the seafloor at 50 m depth that form a dense array of 16 km^2 in the center of the domain. In the fluid domain, we set $\rho_F = 1{,}000\,\text{kg/m}^3$ and $c = 1{,}500\,\text{m/s}$. In the solid domain, we assume a constant density of 2,300 kg/m^3 and a constant Poisson's ratio of 0.25, i.e. we have the relation $v_p = \sqrt{3} v_s$ for the velocity of compressional and shear waves and only

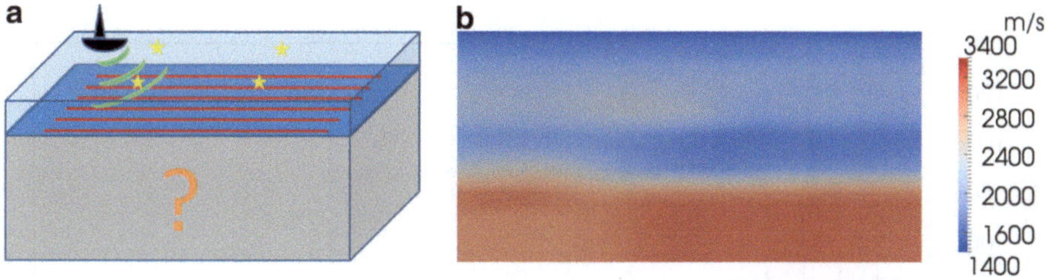

Fig. 3 (**a**) Data acquisition in marine geophysical exploration. A research vessel equipped with an air gun cruises over the field of interest and emits pressure waves into the sea (*green*). An array of geophones is buried in the shallow sea floor (*red lines*) to record seismic waves that are reflected from the subsurface. Based on these measurements, the material structure in the solid domain (*gray*) shall be reconstructed. (**b**) shows the P-wave velocity of the target model for a vertical cross section through the domain

one parameter field to invert for. The synthetic target model has P-wave velocities that range from 1,400 to 3,400 m/s. The data is generated by running a simulation with the target model. In order to deal with the difficulties in using synthetic data and their potential for committing inverse crimes [22], we use a finer mesh for both, the parameter and the state space and add 2 % Gaussian noise to the seismograms. The reference model varies only vertically and for every fixed depth, we use the average value of the target model in the horizontal plane as the reference value.

Since the computational domain is artificially truncated, we have to impose absorbing boundary conditions to reduce reflections from non-physical boundaries. We follow ideas from [23] and apply dampers that relate the traction to the velocity in the solid domain and a Sommerfeld-like condition in the fluid domain:

$$\nabla \chi \cdot \mathbf{n} = -c^{-1} \chi_t \qquad \text{on } \Gamma_F^{\text{abs}} \times I, \tag{23}$$

$$(\Psi : \varepsilon(u)) \cdot \mathbf{n} = v_p \rho_S \, (u_t \cdot \mathbf{n}) \, \mathbf{n} + v_s \rho_S \, (u_t - (u_t \cdot \mathbf{n}) \, \mathbf{n}) \qquad \text{on } \Gamma_S^{\text{abs}} \times I. \tag{24}$$

Furthermore, we enforce $\Phi(m) = 0$ on Γ_S^{abs}, i.e. the parameter model is not updated on the artificial boundaries, in order to avoid artifacts in the reconstruction.

We solve the seismic inverse problem by sequentially inverting for source frequencies of 0.625, 1.25 and 2.5 Hz. On the finest level, we obtain a parameter mesh with approx. 100k degrees of freedom. The state mesh has approx. 300k spatial grid points and 1,000 time steps.

Figure 4 shows histograms for the misfit at all receiver locations before and after the optimization. The accumulated misfit has been reduced by more than 83 %, i.e. there is a good match between observed and reconstructed data. Figure 5 compares the initial, target and reconstructed parameter model at a vertical cross section through the domain. The inversion output looks reasonable, especially near the surface. As expected the reconstruction becomes less accurate at greater depths. Figure 6 shows horizontal snapshots of the P-wave velocity for the reconstructed and the target model. The adaptively refined parameter meshes are illustrated in Fig. 7.

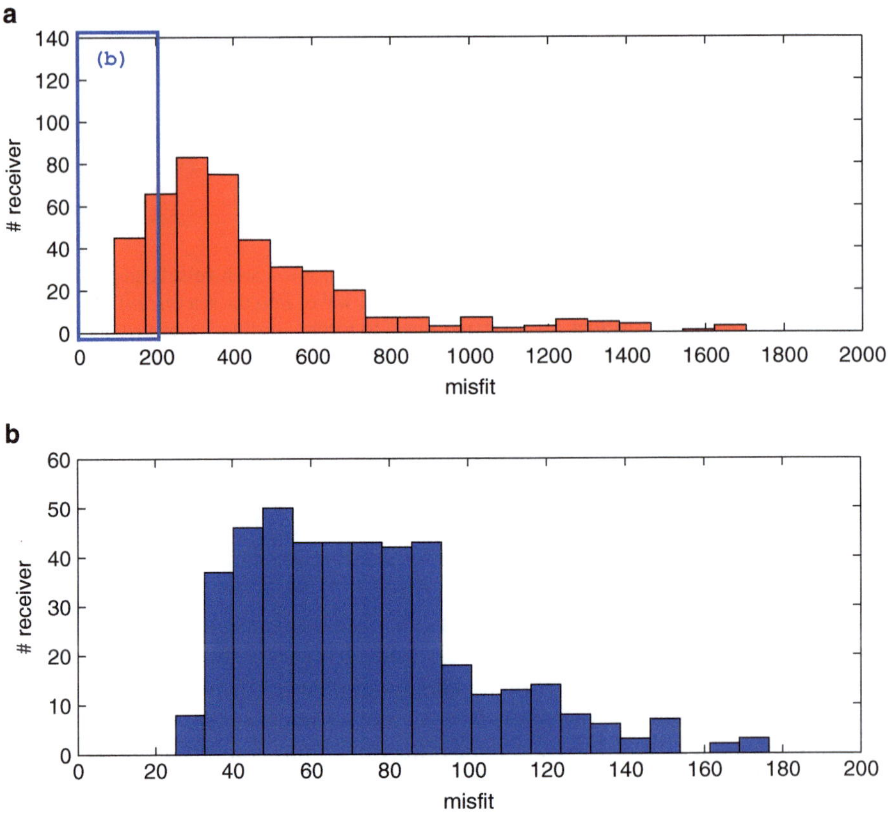

Fig. 4 (**a**) Histogram of the misfit at all receiver locations using the initial model. (**b**) shows the histogram obtained with the reconstructed model. Note the different scaling of the x-axis. The misfit has been reduced by more than 83 %. All three components of the seismograms are used to compute the misfit

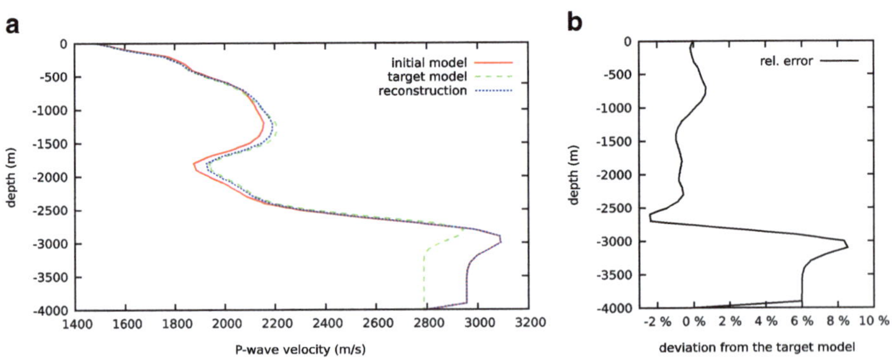

Fig. 5 (**a**) P-wave velocity for a vertical cross section through the domain at the center of the x-y-plane. (**b**) Relative error between the reconstruction and the target model

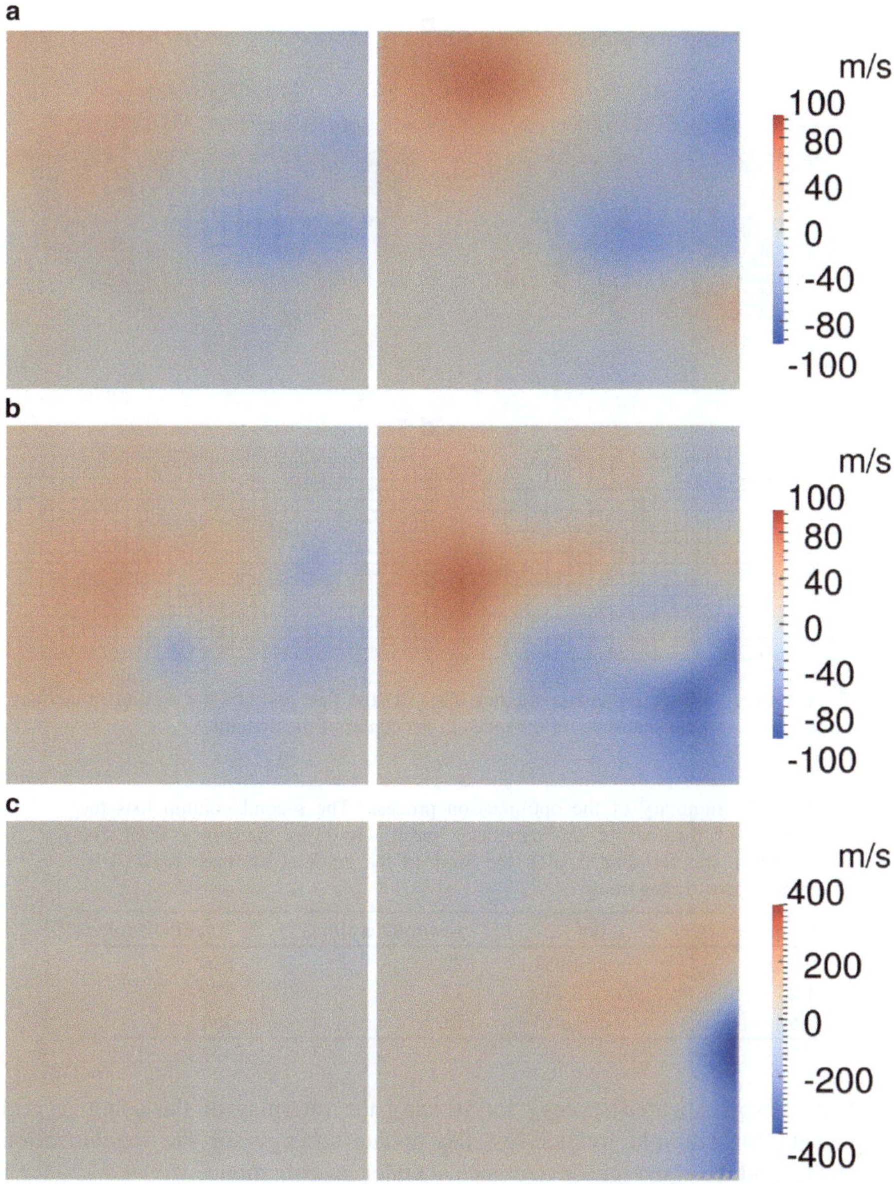

Fig. 6 Horizontal snapshots of the reconstructed (*left*) and target (*right*) v_P at (**a**) 150 m, (**b**) 800 m and (**c**) 1,500 m depth. The images show the deviation from the reference model that is homogeneous for every fixed depth

Table 2 summarizes the optimization process on the different frequency levels. The stopping criterion was a relative reduction of the norm of the gradient by three orders of magnitude. The computations were carried out on a Linux cluster using 32 processors.

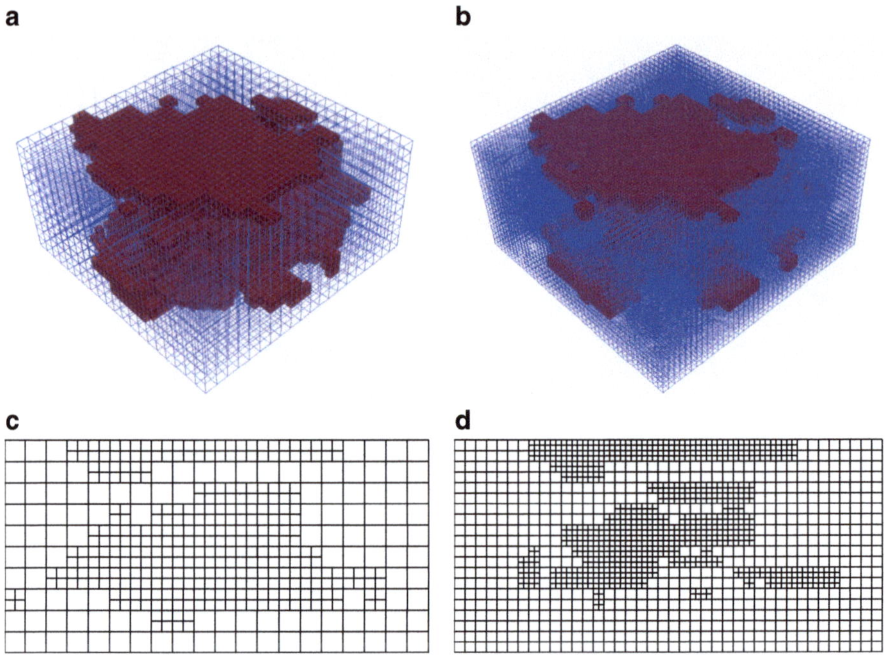

Fig. 7 Adaptively refined parameter meshes after (**a**) the first and (**b**) the second refinement. (**c**) and (**d**) show vertical snapshots of the mesh in the center of the domain

Table 2 Summary of the optimization process. The second column lists the degrees of freedom of the parameter mesh. On every frequency level, the algorithm was terminated, after the norm of the gradient has been reduced by three orders of magnitude

Frequency (Hz)	dof	Newton iterations	Total PDE solves
0.625	4,851	22	886
1.25	15,949	15	550
2.5	101,015	13	384

A more sophisticated strategy for steering the interplay of the adaptive grid refinement for multiple frequencies, the update strategy for the regularization parameter and the convergence tolerance is under investigation.

5 Conclusion

We have presented a Newton-CG method for full-waveform seismic tomography in a coupled solid-fluid system. The optimization framework consists of a trust-region globalization, a smooth cutoff function and a multi-frequency inversion approach.

Derivatives are efficiently computed using adjoint techniques. Numerical results show the applicability of our method to problems in marine geophysical exploration. A couple of major extensions in both, theoretical and practical aspects, will appear in [5].

The fast local convergence of the proposed method is superior over approaches that do not incorporate second order information. Furthermore, in the context of statistical inverse problems, the Hessian of the cost functional can be used to approximate the a-posteriori density of the parameter model [17, 26]. On the other hand, the Newton-type method requires a high number of simulations per iteration in comparison to purely gradient-based or limited-memory BFGS methods. In our implementation, we limit this extra effort by terminating the CG iterations early. Additionally, the multi-frequency inversion approach allows to perform several iterations on a coarser mesh at significantly lower costs and the starting model on the finest grid can be expected to be already in the vicinity of the minimizer and to require fewer Newton iterations.

Future work will incorporate randomized data reduction techniques and a sample average approximation [19]. We will also improve the parallel efficiency of the simulation code.

Acknowledgements The project is supported by the Munich Centre of Advanced Computing, Technische Universität München, Germany. The computations were carried out on a Linux cluster that is partially funded by the grant DFG INST 95/919-1 FUGG. In addition, the first author gratefully acknowledges support by the International Graduate School of Science and Engineering at the Technische Universität München, Germany. The authors would like to thank Heiner Igel, Alan Schiemenz and three anonymous referees for helpful comments and valuable suggestions. We would also like to thank Michael Bader for initiating and editing this volume.

References

1. Aravkin, A., Friedlander, M., Herrmann, F., van Leeuwen, T.: Robust inversion, dimensionality reduction, and randomized sampling. Math. Program. **134**, 101–125 (2012)
2. Bangerth, W., Kanschat, G.: deal.II Differential Equations Analysis Library, Technical Reference (2013). http://www.dealii.org
3. Bangerth, W., Geiger, M., Rannacher, R.: Adaptive Galerkin finite element methods for the wave equation. Comput. Methods Appl. Math **10**(1), 3–48 (2010)
4. Becker, R., Rannacher, R.: An optimal control approach to a posteriori error estimation in finite element methods. Acta Numer. **10**(1), 1–102 (2001)
5. Boehm, C., Ulbrich, M.: An adaptive semismooth Newton-cg method for constrained parameter identification in seismic tomography (2013, Preprint)
6. Brossier, R., Operto, S., Virieux, J.: Seismic imaging of complex onshore structures by 2D elastic frequency-domain full-waveform inversion. Geophysics **74**(6), WCC105–WCC118 (2009)
7. Bunks, C., Saleck, F., Zaleski, S., Chavent, G.: Multiscale seismic waveform inversion. Geophysics **60**(5), 1457–1473 (1995)
8. Burstedde, C., Ghattas, O.: Algorithmic strategies for full waveform inversion: 1D experiments. Geophysics **74**(6), WCC37–WCC46 (2009)

9. Chaljub, E., Valette, B.: Spectral element modelling of three-dimensional wave propagation in a self-gravitating earth with an arbitrarily stratified outer core. Geophys. J. Int. **158**(1), 131–141 (2004)
10. Conn, A.R., Gould, N.I.M., Toint, P.L.: Trust Region Methods. SIAM, Philadelphia (2000)
11. Dziewonski, A., Anderson, D.: Preliminary reference earth model. Phys. Earth Planet. Inter. **25**(4), 297–356 (1981)
12. Epanomeritakis, I., Akçelik, V., Ghattas, O., Bielak, J.: A Newton-cg method for large-scale three-dimensional elastic full-waveform seismic inversion. Inverse Prob. **24**(3), 034015 (2008)
13. Fichtner, A.: Ses3d version 2.1: programme description and mathematical background. Technical report, Ludwig-Maximilians-Universität München, München (2009)
14. Fichtner, A., Kennett, B., Igel, H., Bunge, H.: Theoretical background for continental- and global-scale full-waveform inversion in the time–frequency domain. Geophys. J. Int. **175**(2), 665–685 (2008)
15. Fichtner, A., Kennett, B., Igel, H., Bunge, H.: Full seismic waveform tomography for upper-mantle structure in the Australasian region using adjoint methods. Geophys. J. Int. **179**(3), 1703–1725 (2009)
16. Fichtner, A., Bleidinhaus, F., Capdeville, Y.: Full Seismic Waveform Modelling and Inversion. Springer, Berlin/Heidelberg (2011)
17. Flath, H., Wilcox, L., Akçelik, V., Hill, J., van Bloemen Waanders, B., Ghattas, O.: Fast algorithms for Bayesian uncertainty quantification in large-scale linear inverse problems based on low-rank partial Hessian approximations. SIAM J. Sci. Comput. **33**(1), 407–432 (2011)
18. Griewank, A.: Achieving logarithmic growth of temporal and spatial complexity in reverse automatic differentiation. Optim. Methods Softw. **1**(1), 35–54 (1992)
19. Haber, E., Chung, M., Herrmann, F.: An effective method for parameter estimation with PDE constraints with multiple right-hand sides. SIAM J. Optim. **22**(3), 739–757 (2012)
20. Heroux, M., Willenbring, J., Heaphy, R.: Trilinos developers guide. Sandia National Laboratories, SAND2003-1898 (2003)
21. Hinze, M., Pinnau, R., Ulbrich, M., Ulbrich, S.: Optimization with PDE Constraints. Springer, New York (2008)
22. Kaipio, J., Somersalo, E.: Statistical and Computational Inverse Problems, vol. 160. Springer, New York (2004)
23. Komatitsch, D., Vilotte, J.P.: The spectral element method: an efficient tool to simulate the seismic response of 2D and 3D geological structures. Bull. Seismol. Soc. Am. **88**(2), 368–392 (1998)
24. Krebs, J.R., Anderson, J.E., Hinkley, D., Neelamani, R., Lee, S., Baumstein, A., Lacasse, M.D.: Fast full-wavefield seismic inversion using encoded sources. Geophysics **74**(6), WCC177–WCC188 (2009)
25. Kröner, A.: Adaptive finite element methods for optimal control of second order hyperbolic equations. Comput. Methods Appl. Math. **11**(2), 214–240 (2011)
26. Martin, J., Wilcox, L., Burstedde, C., Ghattas, O.: A stochastic Newton MCMC method for large-scale statistical inverse problems with application to seismic inversion. SIAM J. Sci. Comput. **34**(3), A1460–A1487 (2012)
27. Meidner, D., Vexler, B.: Adaptive space-time finite element methods for parabolic optimization problems. SIAM J. Control Optim. **46**(1), 116–142 (2007)
28. Nissen-Meyer, T., Fournier, A., Dahlen, F.A.: A 2-D spectral-element method for computing spherical-earth seismograms – ii. Waves in solid-fluid media. Geophys. J. Int. **174**(3), 873–888 (2008)
29. Nolet, G.: A Breviary of Seismic Tomography. Cambridge University Press, Cambridge (2008)
30. Peter, D., Komatitsch, D., Luo, Y., Martin, R., Le Goff, N., Casarotti, E., Le Loher, P., Magnoni, F., Liu, Q., Blitz, C., Nissen-Meyer, T., Basini, P., Tromp, J.: Forward and adjoint simulations of seismic wave propagation on fully unstructured hexahedral meshes. Geophys. J. Int. **186**(2), 721–739 (2011)
31. Plessix, R.E.: A review of the adjoint-state method for computing the gradient of a functional with geophysical applications. Geophys. J. Int. **167**(2), 495–503 (2006)

32. Pratt, G., Shin, C., Hicks, G.J.: Gauss-Newton and full Newton methods in frequency-space seismic waveform inversion. Geophys. J. Int. **133**(2), 341–362 (1998)
33. Prieux, V., Brossier, R., Gholami, Y., Operto, S., Virieux, J., Barkved, O.I., Kommedal, J.H.: On the footprint of anisotropy on isotropic full waveform inversion: the Valhall case study. Geophys. J. Int. **187**(3), 1495–1515 (2011)
34. Rawlinson, N., Pozgay, S., Fishwick, S.: Seismic tomography: a window into deep earth. Phys. Earth Planet. Inter. **178**(3–4), 101–135 (2010)
35. Sirgue, L., Barkved, O., Dellinger, J., Etgen, J., Albertin, U., Kommedal, J.: Full waveform inversion: the next leap forward in imaging at Valhall. First Break **28**, 65–70 (2010)
36. Steihaug, T.: The conjugate gradient method and trust regions in large scale optimization. SIAM J. Numer. Anal. **20**(3), 626–637 (1983)
37. Tape, C., Liu, Q., Maggi, A., Tromp, J.: Seismic tomography of the southern California crust based on spectral-element and adjoint methods. Geophys. J. Int. **180**(1), 433–462 (2010)
38. Tromp, J., Tape, C., Liu, Q.: Seismic tomography, adjoint methods, time reversal and banana-doughnut kernels. Geophys. J. Int. **160**(1), 195–216 (2005)
39. Tromp, J., Komatitsch, D., Liu, Q.: Spectral-element and adjoint methods in seismology. Commun. Comput. Phys. **3**(1), 1–32 (2008)
40. Ulbrich, M.: Semismooth Newton Methods for Variational Inequalities and Constrained Optimization Problems in Function Spaces. SIAM, Philadelphia (2011)
41. Virieux, J., Operto, S.: An overview of full-waveform inversion in exploration geophysics. Geophysics **74**(6), WCC1–WCC26 (2009)
42. Wilcox, L.C., Stadler, G., Burstedde, C., Ghattas, O.: A high-order discontinuous Galerkin method for wave propagation through coupled elastic-acoustic media. J. Comput. Phys. **229**(24), 9373–9396 (2010)

Advances in the Parallelisation of Software for Quantum Chemistry Applications

Martin Roderus, Alexei Matveev, Hans-Joachim Bungartz, and Notker Rösch

Abstract Density functional theory (DFT) provides some of the most important methods used in computational theory today. They allow one to determine the electronic structure of finite chemical systems, be they molecules or clusters, using a quantum-mechanical model, and exposes, thus, the great majority of the systems' properties relevant to chemical applications. However, the numerical treatment of large chemical systems proves to be expensive, requiring elaborate parallelisation strategies.

This paper presents two recent developments which aim at improving the parallel scalability of the quantum chemistry code ParaGauss. First, we introduce a new Fortran interface to parallel matrix algebra and its library implementation. This interface specifies a set of distributed data objects, combined with a set of linear algebra operators. Thus, complicated algebraic expressions can be expressed efficiently in pseudo-mathematical notation, while the numerical computations are carried out by back-end parallel routines. This technique is evaluated on relativistic transformations, as implemented in ParaGauss.

The second development addresses the solution of the generalized matrix eigenvalue problem—an inherent step in electronic structure calculations. In the case the

M. Roderus (✉) · H.-J. Bungartz
Department of Informatics, Technische Universität München, 85748 Garching, Germany
e-mail: roderus@in.tum.de; bungartz@in.tum.de

A. Matveev
Department Chemie, Technische Universität München, 85747 Garching, Germany
e-mail: matveev@theochem.tu-muenchen.de

N. Rösch
Department Chemie & Catalysis Research Center, Technische Universität München, 85747 Garching, Germany

Institute of High Performance Computing, Agency of Science, Technology, and Research, 138632, Singapore
e-mail: roesch@mytum.de; roesch@ihpc.a-star.edu.sg

M. Bader et al. (eds.), *Advanced Computing*, Lecture Notes in Computational Science and Engineering 93, DOI 10.1007/978-3-642-38762-3_6,

symmetry of a molecule is exploited, pertinent matrices expose a block-diagonal structure which makes the efficient use of existing parallel eigenvalue solvers difficult. We discuss a technique that uses a malleable parallel task scheduling (MPTS) algorithm for scheduling instances of parallel ScaLAPACK-routines on the available processor resources. This technique significantly improves the parallel performance of this numerical step, reducing the corresponding execution time to below 1 s in most applications considered.

Keywords High performance computing • Parallel numerical algebra • Density functional theory • Relativistic quantum chemistry • Scheduling algorithms

1 Introduction

Many current implementations of first-principles electronic structure methods for molecules, clusters, and local models of surfaces and solids are limited in their parallel scalability due to the intrinsic diversity of structures and algorithms involved in various subtasks. We refer here to methods that aim at exploiting advantages of expansion techniques based on localized (Gaussian-type) functions. The inherent structure of the iterative self-consistent field (SCF) problem does not admit a homogeneous parallelisation strategy, a fact that severely limits current opportunities in modelling complex chemical systems like catalysts, nano-structured materials as well as large complexes in solution. This paper discusses two recent developments, which address this problem, and provide parallelisation strategies for numerical problems that appear specifically in electronic structure codes. The presented approaches are accompanied by library implementations that are invoked by the Gaussian-based density-functional code ParaGauss [4].

Scientific codes often contain linear algebra expressions, systems of linear equations, and eigenvalue problems. An example is the relativistic transformations and solving the block-diagonal eigenvalue problem for the resulting Hamiltonian as implemented in the quantum chemistry program package ParaGauss [4]. An abstraction for linear algebra operations facilitates a *separation of concerns*, where mathematics is separated from details of the technical implementation. This allows a quick implementation of matrix and vector transformations, thus contributing to the productivity of developing numerical routines. Furthermore, as these transformations may have the appearance of mathematical expressions, the application semantics is easily comprehensible, hence improving the readability of the code.

A number of languages or library extensions exist which provide such a functionality: *Matlab* and *Octave* are high-level scripting languages which operate only with mathematical objects, but are usually not suitable for high-performance codes and large software projects. The C++ libraries *Armadillo* [27], *uBLAS* (as part of *BOOST* [19]) and *MTL* [13], among others, offer template-based matrix classes with comprehensive functionality, and partially also advanced linear algebra operations, such as factorisations or eigenvalue problems. *Fortran* provides an intrinsic programming model, which allows to formulate basic matrix algebra in the source code, using variables and one- and two-dimensional arrays as representations for scalars, vectors and matrices, respectively. The *Matran* [29] library yields further matrix functionality for Fortran, together with advanced operations. However, except for a commercial version of the MTL (*Supercomputing Edition*), software or literature about abstractions supporting parallelism and data distribution, especially for Fortran, is scarce.

In data-intensive applications, it is often desired to distribute the underlying arrays over compute nodes, to avoid memory bottlenecks. Furthermore, expensive operations, especially BLAS Level 3 and eigenvalue computations, are often executed by parallel routines. However, common available parallelisation techniques, such as MPI or Co-Arrays in Fortran 2008, provide a rather low-level interface to express data distribution, and available parallel numerical routines, such as PBLAS and ScaLAPACK [5], work with specific distributed data types and do not allow to express parallel matrix algebra in mathematical notation, as described above. The PLAPACK package [31] addresses this issue and defines more object-style distributed data types along with automatic distribution routines, allowing thus a more modern programming style. However, even though the APIs of those linear algebra libraries are carefully designed, the routine signatures are still complicated, making elaborate algebraic code tedious to write. Hence, separation of concerns is difficult to comply with, which can easily lead to poor code quality.

In Sect. 2 we review a Fortran interface to linear algebra routines and its implementation, as introduced in Ref. [25]. The interface specifies distributed data types, representing matrix operands, and overloads the existing operators $\{+, -, *, **\}$ to accept them. We show how this combination makes it is possible to express matrix algebra in clear mathematical notation directly in the code, while operating with distributed data and parallel routines at its back-end. We demonstrate how the newly established interface can be used to parallelise an existing sequential implementation of relativistic transformations in a quantum chemistry code and include a few benchmarks.

Furthermore, in Sect. 3 we present an approach to solve the generalized matrix eigenvalue problem

$$HC = SCE \tag{1}$$

with symmetric block-diagonal matrices H and $S > 0$. The matrix E is diagonal, and contains the eigenvalues, and the columns of C the corresponding eigenvectors. Equation (1) for dense symmetric matrices can be solved by existing parallel solvers, provided e.g. by the libraries ScaLAPACK [5], PLAPACK [31] or ELPA [2]. However, a more complicated situation arises when the spatial symmetry of a molecule or cluster is exploited to reduce the dense eigenvalue problem to a few subproblems of smaller dimensions [24].

While the majority of current quantum chemistry problems are solved without invoking spatial symmetry constraints, exploiting point group symmetry can offer significant advantages in the field of nanoparticles [16, 26]. Clusters of various materials in the size range of nanoparticles (1–10 nm diameter) typically contain about 50 to several thousand atoms. Molecules with more than 100 heavy atoms still represent a challenge to accurate quantum chemistry methods and symmetry thus yields a convenient way of applying these methods to nano-sized model species in the relevant size range. Such nanoparticles represent prototypical applications, where a generalized matrix eigenvalue problem, decomposed into several sub-matrices (typically 4–10) of different sizes, has to be solved. This simplification significantly reduces the computational effort; however the implied block-diagonal matrix structure cannot be handled efficiently by existing (parallel) eigenvalue solvers. An approach to solving this problem is discussed in Sect. 3.

2 A Fortran Interface to Parallel Matrix Algebra

This section presents the specification and implementation of a set of high-level Fortran data types, functions, subroutines, and operator interfaces, which facilitate the handling of distributed data objects as a representation of matrix operands, as well as parallel arithmetic operations on them. Basic linear algebra can be expressed directly in common mathematical notation, more advanced operations, such as eigenvalue computations and domain specific operations, by subroutine calls. We have targeted and achieved a programming model which allows a separation of concerns: mathematical expressions are stated as simple as possible in the source code, while the library takes care of the low-level parallelisation, reliability and performance. As Sect. 2.3 shows, the library has been tested, and integrated into the quantum chemistry software ParaGauss [4].

In addition, the following requirements were considered in the interface design: (i) easy data object management (creation, disposal of, etc.) with technical details, such as physical data distribution, being hidden from the user, and (ii) an interface implementable with Fortran, MPI, and performance-optimized external libraries.

The interface realizes a *single instruction multiple data* (SIMD) programming model, in contrast to the *single program multiple data* (SPMD) style, often applied in pure MPI codes. Here, each library call represents a collective operation, encouraging synchronous control flow. The resulting source code looks like a sequential program, parallelism is achieved by operations on distributed objects by library routines. The internals of the opaque distributed objects, including the actual numeric data, are not exposed. These design features entail the advantage of good readability. Furthermore, adoption of serial code basically requires a change of the affected data types, as well as invocation of the library with a USE-statement.

The user is given the opportunity to choose the processes, involved in the data distribution and computation, via a communication context. This allows combining different paradigms and parallelisation technologies—the user is not restricted to a single paradigm for the overall application.

We defined two data types representing distributed dense and diagonal matrices: rmatrix and rdmatrix. The generic constructor MATRIX converts a 1- or 2-dimensional array into the distributed matrix object [25]. The complementary

```
1 real(DP) :: A(n), B(n, n), C(n, n)
2 type(rdmatrix) :: mA
3 type(rmatrix) :: mB, mC
4 ...
5 mA = matrix(A) ! Create distributed objects
6 mB = matrix(B)
7
8 mC = mA * mB   ! Matrix arithmetics
9
10 C = array(mC) ! Make an array from distributed data
```

Fig. 1 Demonstration of the API usage. Line 1 declares arrays whereas lines 2–3 declare two types of distributed matrices. In lines 5–6, distributed data objects for the diagonal and the dense matrix, respectively, are created. Line 8 shows a simple arithmetic operation. Finally, in line 10 the computed dense matrix is stored in a plain array. In this particular case the current implementation requires communication only in line 10

generic function ARRAY converts a matrix object back into a plain array. Furthermore, we overload some of Fortran intrinsic unary and binary arithmetic operators to express parallel operations between the distributed data types. These operators include: binary addition, binary subtraction, unary negation, binary (matrix) multiplication, exponentiation. Matrix transposition is implemented as a unary function tr(). The generic constructors MATRIX and ARRAY are sufficient to implement arbitrary matrix functions. For example, a reference implementation for the function tr() is a composition of Fortran intrinsic function, array and matrix constructors: tr(A) = matrix(transpose(array(A))). Figure 1 shows a simple code example, which demonstrates the general usage of the API presented here.

2.1 Interface Implementation

The API has been implemented as a library with Fortran bindings. The implementation language is Fortran 95 with the *enhanced data type facilities* (EDF), documented in the technical report TR15581 [11]. For our purposes, allocatable components, documented in this report, provide automatic deallocation upon object destruction and the "move semantics" similar to that of "rvalues" formalized in the C++11 standard. Additionally, we rely on MPI for communication primitives, and external, parallel library routines for expensive operations: PDGEMM from the PBLAS library for the dense matrix multiplication, PDSYGVX and PDTRAN from the ScaLAPACK library for the dense generalized eigenvalue problem, and the matrix transposition, respectively [5].

The matrix operands are represented by data types, containing a (distributed) allocatable array, and meta data, necessary for MPI, PBLAS, and ScaLAPACK usage, see Fig. 2. The arrays representing dense matrices are distributed in a block-cyclic manner, a native distribution scheme of PBLAS and ScaLAPACK (see the ScaLAPACK user's guide [5]). With this data distribution scheme, the array

```
type, public :: rmatrix
    private
    integer :: mpi_comm = MPI_COMM_NULL, desc(9) = -1
    real(DP), allocatable :: m(:, :)
end type rmatrix

type, public :: rdmatrix
    private
    integer :: mpi_comm = MPI_COMM_NULL, blacs_ctxt = -1
    real(DP), allocatable :: d(:)
end type rdmatrix
```

Fig. 2 Declaration of the opaque rmatrix and rdmatrix data types, in Fortran notation. An integer array of length 9 is used by ScaLAPACK for a matrix descriptor and holds, among other data, matrix dimensions and a BLACS communicator

```
interface operator(*)
    module procedure mult_r_r ! dense * dense
    module procedure mult_d_d ! diagonal * diagonal
    module procedure mult_r_d ! dense * diagonal
    module procedure mult_d_r ! diagonal * dense
    module procedure mult_r_0 ! dense * scalar
    ...
end interface operator(*)
```

Fig. 3 Multiplication Fortran operator interface. The expression A * B * C for dense matrices A and B and diagonal matrix C will resolve to mult_r_d(mult_r_r(A, B), C)

constructor instance array_r of the generic ARRAY function, assembling a full 2D array on each process, is the only operation that requires communication [25].

For arithmetic operations between these data types, we overload the existing intrinsic operators with the interface construct. Figure 3 demonstrates this for the multiplication operator *. Here, the compiler does a static type check and resolves an operation to a specific function call, see Fig. 3. The storage for the function result is explicitly allocated in the function body. For the automatic deallocation of intermediate results, we rely on the allocatable semantics of the EDF, as explained at the beginning of this subsection. This important feature facilitates arithmetic expressions without introducing memory leaks and, together with the move semantics, allows an optimizing Fortran compiler to avoid unnecessary copying of data from the right-hand side of a matrix-valued expression in an assignment statement.

The diagonal matrix, represented by the data type rdmatrix, is stored in the 1D array d(:), see Fig. 2. As the storage requirement for a diagonal matrix is far less demanding than that for a dense matrix, we replicate the data on every process. This significantly simplifies implementation of operations involving diagonal matrices, and saves communication overhead at the cost of slightly higher memory consumption.

2.2 The Test Case: Background

We evaluated the approach to matrix algebra just described with a test case in the context of the quantum-chemistry code ParaGauss [4], which inspired us to develop this library abstraction. The test case covers dense matrix algebra, multiplications of dense and diagonal matrices, generalized eigenvalue problem and some domain specific operations on matrices. This section describes the background of the test case in abstract mathematical notation. In the next section we will show how this formalism translates into source code, and present some benchmarks.

Relativistic quantum chemistry covers a variety of approximations that account for the fact that the velocities of (core) electrons in heavy atoms is sufficiently close to the speed of light. A popular relativistic approximation is derived from the Douglas–Kroll approach [10] enabled by a practical scheme of building matrix representations of integro-differential operators [8]. The transformation starts with the four dense matrices, T, S, V, and O, comprising matrix representations of kinetic energy T, overlap S of basis functions, and matrix representations of two potential terms, V and O [8, 10]. The output consists of the relativistic counterparts of the kinetic energy T_{rel} and potential matrix V_{rel}. The first subproblem to address is a generalized eigenvalue problem for the matrices T and S which amounts to finding a matrix U and a diagonal matrix t such that $U^T TU = t$ and $U^T SU = 1$ [8]. The kinetic energy eigenvalues, t, are further used to compute relativistic factors, formally diagonal matrices of the same dimension: $t_{\mathrm{rel}}, e, a, b, r = \mathrm{factors}(2t)$.

The next step of the second-order DKH transformation is to transform the input matrices V and O: $\tilde{V} = U^T VU$, and $\tilde{O} = U^T OU$. The central piece of the transformation is the matrix equation for the relativistic potential:

$$\tilde{V}_{\mathrm{rel}} = a\tilde{V}a + b\tilde{O}b + R^T er^{-2}R + (eR^T r^{-2}R + R^T r^{-2}Re)/2 \qquad (2)$$

where we introduce the intermediate matrix $R = \mathrm{rpt}(e, r^2 a\tilde{V}a - b\tilde{O}b)$ using the matrix valued function, $\mathrm{rpt}(e, X)$, defined as $\mathrm{rpt} : (e, X) \mapsto Y$, $Y_{mn} = X_{mn}/(e_m + e_n)$. Finally, a back-transformation follows: $T_{\mathrm{rel}} = U^{-T} t_{\mathrm{rel}} U^{-1}$, $V_{\mathrm{rel}} = U^{-T} \tilde{V}_{\mathrm{rel}} U^{-1}$.

2.3 Evaluation

Translated into Fortran, the central part of the DKH transformation, Eq. (2), is shown in Fig. 4. The code initiates two dense matrix multiplications, five dense matrix additions and a few multiplications of dense and diagonal matrices. Note that declarations of distributed matrices do not reserve space for actual data. It is the responsibility of the primitive operations that return an rmatrix or an rdmatrix to reserve space, initialize and finally fill those structures with data. Upon return from the subroutine, all intermediate data structures, declared in the scope of

```
subroutine dkh2(t, V, O, t_rel, V_rel)
    use matrix_parallel, only: rmatrix, rdmatrix, operator(*), &
        operator(+), operator(-), operator(**), tr, rpt
    implicit none
    type(rdmatrix), intent(in) :: t
    type(rmatrix), intent(in) :: V, O
    type(rdmatrix), intent(out) :: t_rel
    type(rmatrix), intent(out) :: V_rel
    ! *** end of interface ***

    type(rdmatrix) :: e, a, b, r2
    type(rmatrix) :: aVa, bOb, RW, W22

    call factors(2.0d0 * t, t_rel, e, a, b, r2)

    aVa = a * V * a
    bOb = b * O * b

    RW = rpt(e, r2 * aVa - bOb)

    W22 = tr(RW) * (0.5d0 * r2**(-1)) * RW

    V_rel = aVa + bOb + tr(RW) * (e * r2**(-1)) * RW &
                     + e * W22 + W22 * e
end subroutine dkh2
```

Fig. 4 Fortran code for the popular form of relativistic transformation, Eq. (2), illustrating the use of the matrix algebra API. A function returning more than one matrix, here the diagonal matrix t_{rel} and the dense matrix V_{rel}, is commonly implemented as a subroutine with multiple output arguments. Declarations of distributed matrices do not lead to memory reservation for actual data. This is carried out by functions implementing primitive matrix operations. For local variables, such as e or aVa, these resources are freed on return from the function

the subroutine, will be automatically freed by a standard TR15581 conforming compiler. Apart from the module name in the USE statement, no other modifications were necessary to parallelise the serial version of this code. Imperative coding style with destructive matrix updates (not shown here) is supported as well [25].

Run time performance has been evaluated at the facilities provided by Leibniz Rechenzentrum, Munich, on the migration system SuperMIG, built by IBM [23]. The machine comprises 205 nodes, each hosting four 10-core Intel Xeon Westmere-EX processors. The nodes are equipped with 256 GB of memory and interconnected by an Infiniband QDR network. With this topology calculations involving, for example, 36, 64 and 81 processor cores were scheduled on 1, 2, or 3 nodes, respectively. For the benchmarks, we employed the default vendor-supplied MPI library implementation, IBM MPI v5.2, and BLACS/ScaLAPACK v1.8, linking to the highly optimized BLAS library distributed with Intel's MKL v10.3.

In Fig. 5 we show the scaling behaviour of a dense matrix multiplication as one of the primitive linear algebra operations (black circles). The parallel efficiency

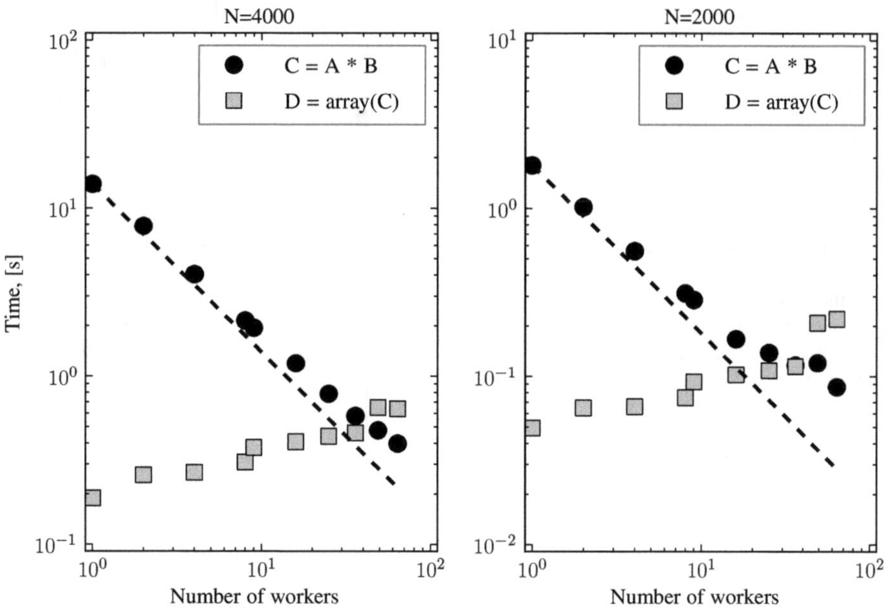

Fig. 5 Log-log plot of the wall-clock time required for a matrix multiplication using PBLAS procedure PDGEMM for matrix dimensions 4,000 (*left pane*) and 2,000 (*right pane*) as a function of the number of cores involved (*black circles*) in comparison to the time required to convert a distributed matrix to a plain array (*gray squares*). Linear regression suggests a power law, $T \sim P^{-\alpha}$, $0.8 < \alpha < 0.9$ for the dependence of matrix multiplication time, T, on core number, P. Ideal scaling, $T \sim P^{-1}$, is shown for comparison (*dashed line*). The times required for matrix multiplication and building a replicated array scale with the matrix dimension N as N^3 and N^2, respectively

is limited by the underlying PBLAS implementation [5]. For a typical matrix dimension of 2,000–4,000, the implementation provided by the local computing centre shows an efficiency above 50 % for a core number below 100. In this example, a linear regression of the data displayed in Fig. 5 suggests a scaling behaviour of the time, T, required by the operation with the number of cores, P, as a power law, $T \sim P^{-\alpha}$ with $\alpha = 0.8 \ldots 0.9$. Note that for these matrix dimensions, the cost of a conversion of a distributed (resulting) matrix to a plain two-dimensional array, replicated on all processes requires an overhead that becomes comparable with the costs of a *single* matrix multiplication at about 40 cores (gray squares, Fig. 5). For higher matrix dimensions, the crossover happens at higher core numbers. Of course, this overhead is amortized when processing more involved linear algebra expressions before, if at all, the results will have to be converted to native arrays.

This is the case for the core of the second-order DKH transformation, see the source code in Fig. 4 implementing Eq. (2), for which both the input and output are distributed matrices. Figure 6 shows how the wall-clock time required for a DKH transformation for two matrix dimensions, $N = 4{,}000$ and $N = 2{,}000$, depends

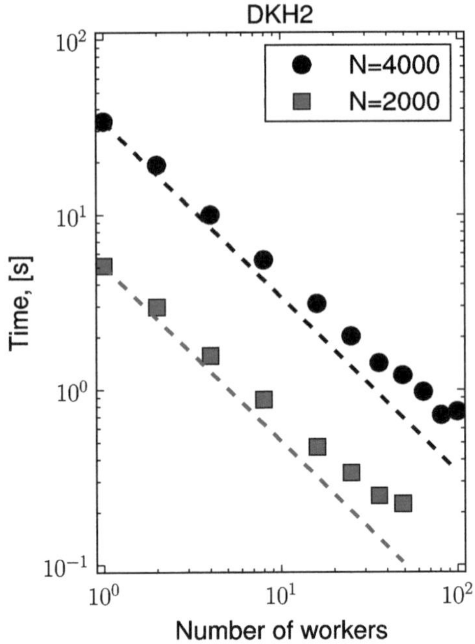

Fig. 6 Log-log plot of the wall-clock time required for the central piece of the second-order DKH transformation for matrix dimensions 4,000 (*black circles*) and 2,000 (*gray squares*) as a function of number of cores. Linear regression suggests that the wall-clock time, T, scales with the number of cores, P, as a power law, $T \sim P^{-\alpha}$, with $0.8 < \alpha < 0.9$. Ideal scaling, $T \sim P^{-1}$, is shown for comparison (*dashed lines*)

on the number of cores used. A linear regression of the data on the log-log scale suggests a power law for the timespan, $T \sim P^{-0.9}$. The parallelisation efficiency is limited primarily by the PBLAS library implementation, cf. Fig. 5. This efficiency approaches 50 % as the number of cores is increased along the horizontal axis. Note that in practical applications, end users very often consider the total timespan as a more relevant metric of the parallelisation efficiency.

3 MPTS-Scheduling of Parallel Eigenvalue Computations

This section discusses a solution to the symmetric, block-diagonal, generalized eigenvalue problem $HC = SCE$, as introduced in Sect. 1. The specific block-diagonal structure of the matrices H and S have the important property that the distinct blocks can be treated as independent subproblems. This property already implies a possibility to introduce parallelism to the diagonalisation step [3]: the sub-matrices are sorted by their size in descending order; whenever a processor becomes available, it is employed to carry out a sequential diagonalisation of the first sub-matrix in the list. However, this approach, also known as *LPT algorithm* [15] as a special form of *list scheduling* [14], shows very limited scalability: the sequential execution time for the largest matrix is a lower bound on the overall execution time, and the number of matrices is an upper bound on the number of processors which can be used. See Fig. 7a.

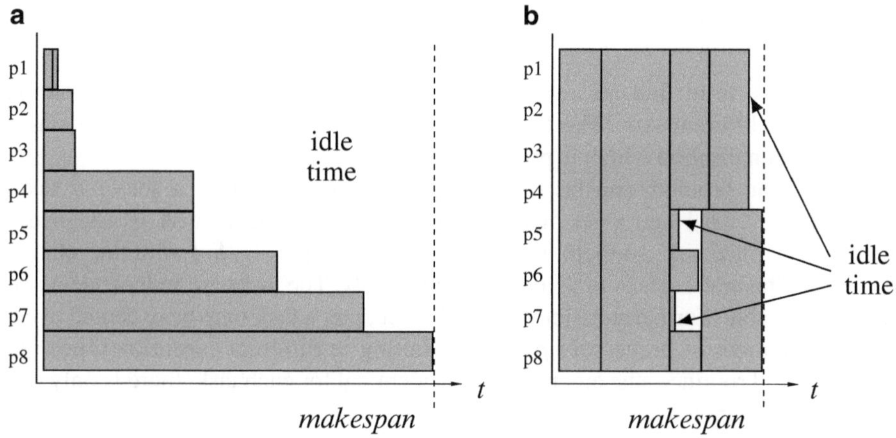

Fig. 7 Example of a scheduling of nine tasks on eight processors. The *grey boxes* represent single tasks. Their widths indicate the execution time and their heights the number of processors they are executed on. (**a**) LPT scheduling. (**b**) Malleable task scheduling

For a typical application, the dimension N of the basis set, which corresponds to the size of H, ranges from 500 up to several thousand. The corresponding generalized eigenvalue problem represents a fraction of the cost of the electronic structure calculation, but it is a fraction growing with $\mathcal{O}(N^3)$ and is intrinsically difficult to parallelise efficiently. Thus, it becomes more important with increasing problem size and CPU count. Moreover, the solution of Eq. (1) is usually part of an iterative scheme with $10^3 - 10^5$ diagonalisations necessary in a single typical application. Thus, an efficient parallel algorithm is required to avoid bottlenecks in large-scale applications.

There exist previous approaches to employ ScaLAPACK [5] as parallel eigensolver in electronic structure calculations, see Refs. [17, 32]. However, in these calculations, the molecular symmetry was not exploited, so H was a single dense, symmetric matrix, which can be treated by standard parallel eigensolvers, such as PDSYEVX or PDSYGVX from ScaLAPACK [5]. In our case, the diagonalisation of $H \in \mathbb{R}^{N \times N}$ can be divided into several smaller sub-problems. Thus, to benefit from these computational advantages, a different parallelisation strategy is required.

Here we describe an approach to tackle this problem, originally reported in Ref. [24]: the sequential and parallel routines DSYGV and PDSYGV from LAPACK [1] and ScaLAPACK [5], respectively, are employed to diagonalise independently the set of sub-matrices. A malleable task scheduling algorithm allots to each matrix a processor count and schedules the instances of the eigensolvers to achieve good load balancing. Thus, the existing parallel eigensolvers can be used efficiently, to reduce the overall execution time.

3.1 Problem Discussion

The objective is to find all eigenvalues and eigenvectors of a set of symmetric matrices of different size. The term "task" will be used as synonym for a single matrix diagonalisation which is processed by an eigensolver.

Hence, the problem can be formulated as follows: given is a set of n tasks $\mathcal{T} = \{T_1, \dots, T_n\}$ and a set of m identical processors. Each task is assigned a matrix of defined size, so there is a set of sizes $\mathcal{S} = \{S_1, \dots S_n\}$ with the relation $T_i \mapsto S_i$. The matrix sizes in $\mathcal{S}$ can vary arbitrarily. The tasks are independent and nonpreemptable. Furthermore they are malleable, i.e., a task may be executed by an arbitrary number of processors $p \leq m$, resulting in different execution times. As the employed routine—the eigensolver—is identical for each task, there is only one *cost function* denoted as

$$t : (S_i, p) \mapsto t_{i,p}, \tag{3}$$

which predicts the execution time of task T_i with size S_i when executed on p processors.

Furthermore, there is a set of processor counts $\mathcal{P}$, where its elements P represent a possible number of processors on which a task can be executed. The *makespan* is the overall time required to process all tasks from $\mathcal{T}$. The goal is to find a scheduling with a minimal makespan. As the problem arises in many practical applications, it is well studied and known as *malleable parallel task scheduling* (MPTS). An example is depicted in Fig. 7b.

Hence, our approach to a solution is as follows: as a first step, we generate an offline scheduling from the task set $\mathcal{T}$ and the process count m, using an MPTS algorithm, see Sect. 3.2. This algorithm requires a cost function, which predicts the time required by a task—here the execution time of the employed eigensolver. Its practical implementation is discussed in Sect. 3.3. This established scheduling is then used in ParaGauss for the efficient parallel computation of the eigenvalue problems.

3.2 Malleable Parallel Task Scheduling

3.2.1 Related Work

The MPTS problem introduced in Sect. 3.1 is a common scheduling problem and was frequently discussed over the last decades, see Refs. [6, 7, 21]. MPTS is a generalisation of a sequential scheduling problem which is NP-complete in the strong sense [12]. Therefore, a variety of approximation algorithms with polynomial runtime exist. A common approach is based on a two-phase strategy, first introduced by Turek et al. [30]. The idea is to find a processor allotment for each task in a first step and to solve the resulting nonmalleable parallel task scheduling (NPTS) problem in a second step. Ludwig and Tiwari [20] suggested such an algorithm with approximation factor 2, which shall be the basis of the strategy proposed here. Mounié et al. [22] followed a different approach by formulating a Knapsack problem as core component. They provide the currently best practical algorithm with approximation factor $\frac{3}{2} + \epsilon$ for any fixed $\epsilon > 0$. When the tasks have to be executed on processors with successive indices, Steinberg [28] proposed an adapted strip-packing algorithm with approximation factor 2. Furthermore, Jansen [18] gave an approximation scheme with makespan at most $1 + \epsilon$ for any fixed $\epsilon > 0$. For special cases of the MPTS, algorithms exist with approximation factors close to one (e.g. Ref. [9] for identical malleable tasks), but those do not apply in our case.

The problem described above is a standard MPTS problem, so the algorithms mentioned could in principle be applied. However, as we will see, the number of sub-matrices to be processed is small due to fundamental symmetry rules. This allows us to modify the algorithm from Ref. [20] by introducing a combinatorial approach in order to find the optimal solution of the NPTS problem.

3.2.2 The Employed Algorithm

As in the approximate algorithm by Ludwig and Tiwari [20], we also follow a two-phase approach. The first phase determines a number of allotted processors for each task using an approximate algorithm where the processor allotment is generated in $\mathcal{O}(mn)$ operations. Thus, the original MPTS problem is transformed into an NPTS problem. In the second phase, a scheduling for the resulting nonmalleable tasks is generated. This can be achieved by applying any (optimal or approximate) NPTS algorithm. An important characteristic of this approach is that the approximation factor of the MPTS problem is equal to the approximation factor of the NPTS algorithm applied in the second phase. Ludwig and Tiwari proposed an algorithm that provides an approximation factor of 2.

In our specific case, the number of tasks is bounded above by 10. This allows an alternative strategy for solving the NPTS problem *exactly* by employing a combinatorial algorithm, different from the algorithm by Ludwig and Tiwari [20]. In our combinatorial approach, the maximum number of possible permutations is $10! \approx 3.6 \cdot 10^6$. However, we also introduced optimisations, which make this worst case very rare. For the detailed algorithm, please refer to Ref. [24].

3.3 *Cost Function*

The scheduling algorithm described requires a cost function which estimates the execution time of the employed eigensolver routines—in our case DSYGV and PDSYGV from the (Sca)LAPACK library. It is difficult to determine upfront how

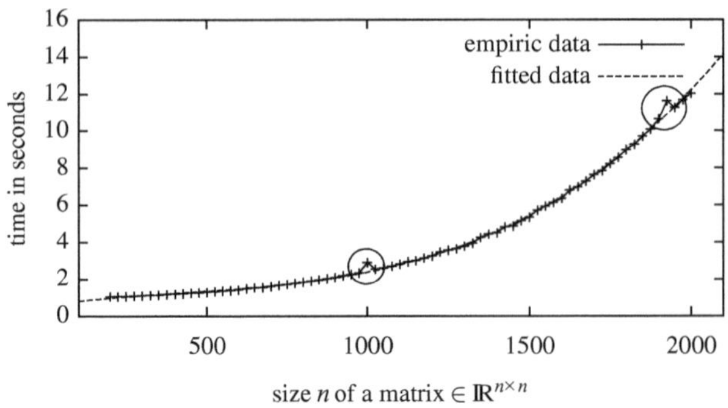

Fig. 8 Execution time measurements of the routine PDSYGV diagonalizing randomly generated matrices. The curve labelled "fitted data" represents a polynomial of degree 3 which was generated by the method of least squares from the empiric data

accurate the estimates have to be. However, the validation of the algorithm will show whether the error bounds are tight enough for the algorithm to work in practice.

There exist analytic models for the performance prediction of parallel routines, see for example the ScaLAPACK User's Guide [5]. However, validation of the models showed that the prediction error usually lies between 10 and 30 %. Furthermore, it does not provide a solution for the practical use in a scheduling algorithm: basically, each routine needs its own model. Therefore, if the routine changes (e.g. due to a revision or the use of a different library), the model has to be adapted as well, which requires in-depth knowledge of the employed routine.

Here we follow a different approach: the routine is handled as a "black box". Predictions of its execution time are based on empirical data which are generated ahead of time on a given computing resource by test runs with a set of randomly generated matrices on a set of possible processor allotments $\mathscr{P}$. Then, with a one-dimensional curve-fitting algorithm, a continuous cost function t is generated for each element of $\mathscr{P}$, Fig. 8. Thus, each $P \in \mathscr{P}$ has a related cost function $t_P : S \mapsto t_{P,S}$, not to be confused with Eq. (3).

Finally, we combine the emerging set of P-related cost functions to form the general cost function, Eq. (3). However, in practice, when a certain number of allotted processors is exceeded, parallel routines no longer feature a speedup or even slow down [32]. This behaviour does not comply with the assumption of a general monotonic cost function. To satisfy this constraint, we define the cost function, Eq. (3), as follows:

$$t_{i,p} = \min_P \{t_{P,S_i} : P \in \mathscr{P} \wedge P \leq p\}. \tag{4}$$

All possible processor counts $P \in \mathscr{P}$ are considered which are smaller than or equal to p. The P which results in the smallest execution time for the given S

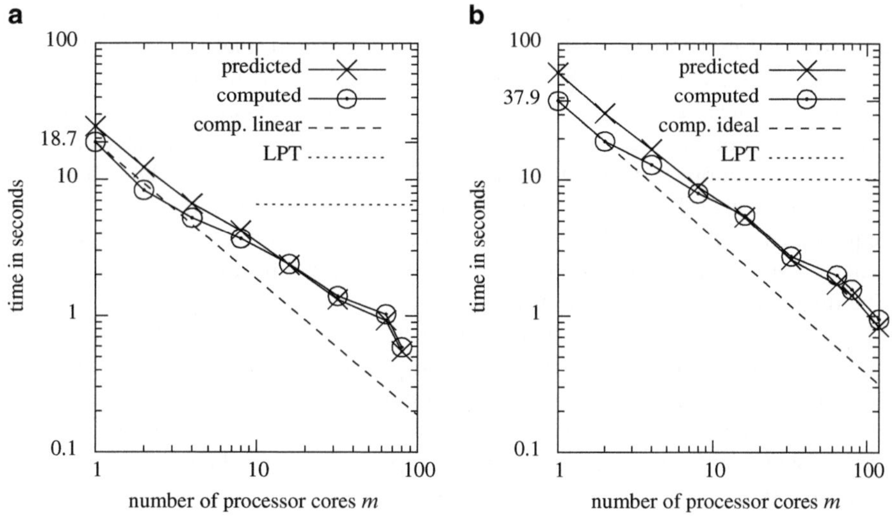

Fig. 9 Log-log time diagrams of the complete eigensolver step of the test systems Pd_{489} and Pd_{670}. Considered is the wall-clock time of the diagonalisation module during one SCF iteration. The curves labelled "predicted" show the makespan of the scheduling algorithm, predicted by the cost function. The curves labelled "computed" provide the real execution time of the scheduled eigensolvers. The lines labelled "LPT" indicate the execution time of the sequential LAPACK routine that computes the largest matrix and yield, thus, the best possible performance of the previously used LPT scheduler. (**a**) Pd_{489}. (**b**) Pd_{670}

also determines the P-related cost function and thus the result t of the general cost function, Eq. (4).

3.4 *Evaluation*

We evaluated the scheduler just described for two molecular systems as example applications: the palladium clusters Pd_{489} and Pd_{670} in symmetry O_h. Both test systems entail 10 matrices, resulting from the applied point group symmetry, with dimensions ranging from 305 to 2,105. On average, the matrix dimensions of the test system Pd_{670} are about 40 % larger than those of the system Pd_{489}.

For the test runs, the migration system SuperMIG, built by IBM and installed at Leibniz Rechenzentrum, was used; see Sect. 2.3 for more details.

We measured the execution times of the complete eigensolver step in a *single* SCF iteration, see Fig. 9. Recall that typical quantum chemical applications require between 10^3 and 10^5 eigenvalue computations. This should be taken into consideration when examining the real-time savings achieved by this parallelisation technique.

Figure 9 shows that the cost function works accurate in most cases, with an error well below 10 %. Interestingly, this does not apply to small processor numbers ($m \leq 2$), where the error is quite significant (up to $\approx$40 % in the Pd_{670} test runs). Fortunately, these cases still provide good scheduling results, even with poorly predicted execution times. For higher processor numbers m, the cost function works sufficiently accurate to facilitate the practical use of the scheduling algorithm.

The figure also shows a lower bound on the execution time of the previously used sequential scheduler ("LPT"-line). As one can see, this performance bound is now broken and the execution time is improved below this limit by our new algorithm. The diagonalisation step of the test system Pd_{489} was sped up by a factor of up to 11.1, compared to LPT, and by a factor of 31.6 compared to a sequential run. For the system Pd_{670} we achieved maximum speed-up factors of 10.5 and 40, compared to LPT and a sequential runs, respectively. For both test systems, the overall execution time of the eigensolver step in one SCF iteration now lies well below 1 s. Thus, we have shown that the execution time of this important step can be reduced down to a minor fraction of the overall electronic structure calculation.

4 Conclusions

We presented a parallel programming interface, which facilitates easy data management and the expression of matrix operations in almost mathematical notation. The evaluation shows that this technique is suitable for real-world problems—in our case relativistic transformations in quantum chemistry software. The resulting code has indeed the appearance of the original abstract mathematical formulation, while tedious, distracting low-level tasks, such as data organisation, are reduced to a few function calls before and after the matrix operations. We furthermore showed that with our implementation, which partly relies on the performance-optimized parallel libraries PBLAS and ScaLAPACK, the relativistic transformations scale up to 81 cores for input matrices of dimension 4,000, requiring less than 1 s for the overall relativistic transformation. This states a major improvement compared to the previous, sequential, implementation. Consequently, this work brings together programming productivity, code quality and parallel performance, which we consider a useful contribution to software engineering in high-performance computing.

Furthermore, a parallel bottleneck of the eigensolver step in ParaGauss, imposed by the previous LPT-based parallelisation approach, could be eliminated. This was achieved by employing a more sophisticated MPTS scheduler, together with parallel eigensolver routines from ScaLAPACK. The overall scalability of the eigensolver step was significantly improved, being now able to use processor core numbers up to 120 efficiently in large-scale chemical applications. The time spent in this step was reduced to a minor fraction of what was necessary for the previous solution, requiring less than 1 s in one SCF iteration for the test cases considered. This approach also goes beyond the use of parallel eigensolvers in other Gaussian-based DFT codes [2, 17]: to our knowledge, it is the first technique that allows an efficient parallel treatment of Hamilton matrices with a block-diagonal structure. Thus, DFT-based methods which achieve computational benefits by exploiting molecular symmetries have now been augmented with a specific, efficient parallelisation technique.

References

1. Anderson, E., Bai, Z., Bischof, C., Demmel, J., Dongarra, J., Du Croz, J., Greenbaum, A., Hammarling, S., McKenney, A., Ostrouchov, S., Sorensen, D.: LAPACK's User's Guide. SIAM, Philadelphia (1992)
2. Auckenthaler, T., Blum, V., Bungartz, H.J., Huckle, T., Johanni, R., Krämer, L., Lang, B., Lederer, H., Willems, P.R.: Parallel solution of partial symmetric eigenvalue problems from electronic structure calculations. Parallel Comput. **37**, 783–794 (2011)
3. Belling, T., Grauschopf, T., Krüger, S., Mayer, M., Nörtemann, F., Staufer, M., Zenger, C., Rösch, N.: In: Bungartz, H.J., Durst, F., Zenger, C. (eds.) High Performance Scientific and Engineering Computing. Lecture Notes in Computational Science and Engineering, vol. 8, p. 439. Springer, Heidelberg (1999)
4. Belling, T., Grauschopf, T., Krüger, S., Nörtemann, F., Staufer, M., Mayer, M., Nasluzov, V.A., Birkenheuer, U., Hu, A., Matveev, A.V., Shor, A.V., Fuchs-Rohr, M.S.K., Neyman, K.M., Ganyushin, D.I., Kerdcharoen, T., Woiterski, A., Gordienko, A.B., Majumder, S., Rösch, N.: PARAGAUSS, version 3.1. Technische Universität München (2006)
5. Blackford, L.S., Choi, J., Cleary, A., D'Azevedo, E., Demmel, J., Dhillon, I., Dongarra, J., Hammarling, S., Henry, G., Petitet, A., Stanley, K., Walker, D., Whaley, R.C.: ScaLAPACK Users' Guide. Society for Industrial and Applied Mathematics, Philadelphia (1997)
6. Blazewicz, J., Kovalyov, M.Y., Machowiak, M., Trystram, D., Weglarz, J.: Scheduling malleable tasks on parallel processors to minimize the makespan. Ann. Oper. Res. **129**, 65–80 (2004)
7. Blazewicz, J., Ecker, K., Pesch, E., Schmidt, G., Weglarz, J.: Handbook on Scheduling: From Theory to Applications. Springer, Heidelberg (2007)
8. Buenker, R.J., Chandra, P., Hess, B.A.: Matrix representation of the relativistic kinetic energy operator: two-component variational procedure for the treatment of many-electron atoms and molecules. Chem. Phys. **84**, 1–9 (1984)
9. Decker, T., Lücking, T., Monien, B.: A 5/4-approximation algorithm for scheduling identical malleable tasks. Theor. Comput. Sci. **361**(2), 226–240 (2006)
10. Douglas, M., Kroll, N.M.: Quantum electrodynamical corrections to the fine structure of helium. Ann. Phys. (NY) **82**, 89 (1974)
11. Enhanced data type facilities. ISO/IEC TR 15581, 2nd edn. (1999). ftp://ftp.nag.co.uk/sc22wg5/N1351-N1400/N1379.pdf
12. Garey, M., Johnson, D.: Computers and Intractability: A Guide to the Theory of NP-Completeness. W.H. Freeman and Company, New York (1979)
13. Gottschling, P., Wise, D.S., Adams, M.D.: Representation-transparent matrix algorithms with scalable performance. In: Proceedings of the 21st Annual International Conference on Supercomputing, ICS '07, Seattle, pp. 116–125. ACM, New York (2007)
14. Graham, R.L.: Bounds for certain multiprocessing anomalies. Bell Syst. Tech. J. **45**, 1563–1581 (1966)
15. Graham, R.L.: Bounds on multiprocessing timing anomalities. SIAM J. Appl. Math. **17**, 263–269 (1969)
16. Häberlen, O.D., Chung, S.C., Stener, M., Rösch, N.: From clusters to the bulk. A relativistic electronic structure investigation on a series of gold clusters Au_n, $n = 6, \ldots 147$. J. Chem. Phys. **106**, 5189–5201 (1997)

17. Hein, J.: Improved parallel performance of SIESTA for the HPCx Phase2 system. Technical report, The University of Edinburgh (2004)
18. Jansen, K.: Scheduling malleable parallel tasks: an asymptotic fully polynomial time approximation scheme. Algorithmica **39**, 59–81 (2004)
19. Ling, B.S.: The Boost C++ Libraries. XML Press (2011). http://books.google.de/books?id=xMH0XwAACAAJ
20. Ludwig, W., Tiwari, P.: Scheduling malleable and nonmalleable parallel tasks. In: SODA '94, Arlington, pp. 167–176 (1994)
21. Mounié, G., Rapine, C., Trystram, D.: Efficient approximation algorithms for scheduling malleable tasks. In: SPAA '99, Saint Malo, pp. 23–32 (1999)
22. Mounié, G., Rapine, C., Trystram, D.: A 3/2-approximation algorithm for scheduling independent monotonic malleable tasks. SIAM J. Comput. **37**(2), 401–412 (2007)
23. National supercomputer HLRB-II. http://www.lrz-muenchen.de/. Retrieved on 10 Aug 2012
24. Roderus, M., Berariu, A., Bungartz, H.J., Krüger, S., Matveev, A.V., Rösch, N.: Scheduling parallel eigenvalue computations in a quantum chemistry code. In: Euro-Par (2)'10, Ischia, pp. 113–124 (2010)
25. Roderus, M., Matveev, A.V., Bungartz, H.J.: A high-level Fortran interface to parallel matrix algebra. In: CCSEIT-2012, International Conference Proceeding Series (ICPS), Coimbatore. ACM (2012, Accepted)
26. Rösch, N., Matveev, A., Nasluzov, V.A., Neyman, K.M., Moskaleva, L., Krüger, S.: Quantum chemistry with the Douglas–Kroll–Hess approach to relativistic density functional theory: efficient methods for molecules and materials. In: Schwerdtfeger, P. (ed.) Relativistic Electronic Structure Theory – Applications. Theoretical and Computational Chemistry Series, vol. 14, pp. 656–722. Elsevier, Amsterdam (2004)
27. Sanderson, C.: Armadillo: an open source C++ linear algebra library for fast prototyping and computationally intensive experiments. Technical report, NICTA (2010)
28. Steinberg, A.: A strip-packing algorithm with absolute performance bound 2. SIAM J. Comput. **26**(2), 401–409 (1997)
29. Stewart, G.W.: Matran: a Fortran 95 matrix wrapper. Technical report, UMIACS (2003)
30. Turek, J., Wolf, J., Yu, P.: Approximate algorithms for scheduling parallelizable tasks. In: SPAA'92, San Diego, pp. 323–332 (1992)
31. van de Geijn, R.A.: Using PLAPACK: Parallel Linear Algebra Package. MIT, Cambridge (1997)
32. Ward, R.C., Bai, Y., Pratt, J.: Performance of parallel eigensolvers on electronic structure calculations II. Technical report, The University of Tennessee (2006)

Designing Spacecraft High Performance Computing Architectures

Fisnik Kraja, Georg Acher, and Arndt Bode

Abstract Recent developments in space applications have indicated that future spacecraft computing platforms will demand for high performance computing (HPC) capabilities. In order to face this challenge, HPC technologies have to be introduced in the design process of such platforms. This paper summarizes some efforts taken to achieve this. A theoretical design for future spacecraft computing platforms is proposed. This design combines traditional reliability techniques and novel HPC solutions for efficient high performance. System components are not specified in terms of type and quantity, but only a logical representation of the system is given. Benchmarking results are obtained on different parallel computing systems to help platform designers in further system specifications. A real space application that reconstructs a synthetic aperture radar (SAR) image is used to benchmark shared memory, distributed memory, and heterogeneous CPU/GPU systems. It turns out that distributed memory systems are a necessity for performance improvements, whereas heterogeneous CPU/GPU systems offer much more efficiency in terms of performance per power consumption, size, and heat dissipation.

Keywords HPC • SAR • Spacecraft computing platform • Heterogeneous CPU/GPU systems • OpenMP • MPI • CUDA

1 Introduction

Space applications are very important in many aspects of our daily life, especially for earth observation and monitoring. Most spacecrafts download collected sensor data to ground centers for further processing. For most applications, raw data size

F. Kraja (✉) · G. Acher · A. Bode
Lehrstuhl für Rechnertechnik und Rechnerorganisation (LRR), Technische Universität München, Munich, Germany
e-mail: kraja@in.tum.de; acher@in.tum.de; bode@in.tum.de

M. Bader et al. (eds.), *Advanced Computing*, Lecture Notes in Computational Science and Engineering 93, DOI 10.1007/978-3-642-38762-3_7,

is much larger than the one of processed data. Taking into consideration that the downlink to earth has a limited bandwidth capacity, it is obvious that downloading raw data takes longer than downloading resulting processed data. Some additional processing time is needed if raw data is downloaded to earth centers for final processing.

The only way to deal with this problem is to improve the on-board processing power so that raw sensor data gets processed on spacecrafts and resulting processed data is then downloaded to earth. With the current technology being used for on-board processing it will be impossible to fulfill computational requirements of future space applications that will demand HPC capabilities. Such applications will have to process more sensor data to produce higher quality results. Most typical applications falling into this category are SAR image reconstruction ones that will have to process huge amounts of data to produce high resolution images and to offer full coverage over time and space [18].

Introducing modern processing technologies for on-board processing in space can be considered not an easy task when thinking of reliability challenges that the radiation environment presents. Other than reliable, these future processing platforms should be portable and scalable to different missions, modular to allow component upgrades and reusability, easily programmable and highly efficient within feasible costs. The only way to achieve these features is to integrate commercial off-the-shelf (COTS) components on a redundant hardware platform and to supplement the system with software implemented fault-tolerance techniques that will offer the required level of reliability.

This paper summarizes the following contributions:

1. The proposal of a high performance computing architecture for future space applications.
2. The implementation of a benchmarking application that uses a spatial frequency interpolation algorithm for SAR image reconstruction.
3. Benchmarking results obtained on novel shared-memory, distributed-memory, and heterogeneous CPU/GPU computing platforms.

2 Multi-core and Many-Core Processors for Space Applications

The fact that commercial processors integrate many cores on a single die can be seen as an opportunity for high performance space computing. Besides, the use of such processors in space can be also seen as a challenge, taking into consideration the radiation environment. The next generation of multi-core processors will increase the number of cores, building so many-core processors. Such processors are the 64-core TILE64 from Tilera [3] and the 80-core tera-scale processor from Intel [10]. By using similar designs for critical applications and on-board science computing, high performance can be obtained by reducing mass and volume. Another novel and interesting architecture is the Plural Architecture [6], which is defined as a shared memory many-core with hardware scheduling. It consists of many small processor cores, each containing a small private memory. A fast network on chip interconnects cores with shared memory, while another network on chip is used to connect them to the scheduler. A 64-core chip with 16 FPUs operating at 400 MHz consumes only 1 W. The Plural many-core is an accelerator single-chip module that runs a single parallel program at a time. This makes it suitable only for solving a specific given problem, but also suitable for space applications, taking into consideration the low-power consumption.

NASA is considering the Maestro many-core processor [19] for reliable space applications. Maestro is a radiation-hardened by-design processor based on the TILE64 processor by Tilera with additional FPUs on each core. Redundancy and fault-recovery techniques have been applied for error detection and permanent recovery. Maestro integrates a grid of 7-by-7 general purpose processing cores (executing at 300 MHz), each composed of a multilevel cache hierarchy with an 8 KB L1 cache and a 64 KB L2 cache. The virtual L3 cache consists of the L2 cache of each tile. This means that it is distributed over all of 49 tiles of Maestro. A 2-dimensional mesh network known as the iMesh [21] interconnects the cores within Maestro. iMesh is responsible for inter-core communication and also for routing data from main memory to individual tiles and I/O interfaces.

Authors in [20] propose new fault-tolerance strategies for the Maestro many-core space-enabled processor because it is still possible to have software and hardware failures, despite the fact that Maestro is radiation-hardened by-design. To minimize the number of such failures, common fault-tolerance strategies used in high performance computing and embedded systems have been applied to Maestro. The available cores and memory controllers offer enough resources for replication, checkpoint and rollback recovery and for application specific fault-tolerance on top of existing radiation-tolerance of Maestro. The purpose of such techniques is to allow the developer to detect and recover from faults within the processor and not to prevent them from occurring. In this way, the computation can continue even in presence of failures, if they do not propagate to the application-level.

Maestro's architecture is very flexible concerning programming models that can be implemented [4]. This flexibility is inherited from the TILE64 processor. Maestro can run an SMP Linux Kernel and supports process- and thread-parallelism. It also supports the iLib [21] task-based parallel programming model from Tilera. Shared memory programming paradigms such as Pthreads and OpenMP can be used for parallelization in Linux. MPI and Tilera's iLib provide message passing abstractions for the distributed memory environment. Commercial Tilera programming models and tools had to be modified mainly to add support for the floating point module included in each of Maestro cores. Even though Maestro is a customized many-core processor based on the commercial TILE64 processor from Tilera, it cannot be considered as a pure COTS product since it was modified in hardware to provide a radiation-hardened by-design processor. There is not yet a final product because Maestro is still in development as part of NASA's OPERA program.

3 The Proposed On-Board HPC Architecture

In order to build a high performance computing platform, multi-core and many-core processor and accelerator technologies have to be introduced in the design process of spacecraft computing platforms. Symmetric and distributed shared memory processor architectures combine a wide range of features to deliver, in a cost effective way, high processing power and reduced bandwidth demands on memories [9]. GPUs have increased their peak performance and bandwidth capacity faster than similarly priced CPUs. Furthermore, GPUs can deliver higher peak computational throughput than CPUs for workloads with abundant parallelism [5]. Using general purpose GPUs and CPUs increases architecture portability and modularity at minimal costs. It also enables flexible platform programming with available parallel programming models like OpenMP [14], MPI [11], and CUDA [13].

To address architecture scalability, a distributed multi-board design is selected. As shown in Fig. 1, a cluster of parallel processing nodes (PPN) that is responsible only for data processing is under the supervision of the radiation hardened management module (RHMU), which takes care of control and management procedures required to improve system reliability. Dataflow in the architecture is from data collecting instruments to PPNs. From there to the backplane and finally to ground via the satellite communication subsystem.

The internal design of each PPN is shown in Fig. 2. I/O digitizer modules are needed to collect data from instruments and to convert it from analog to digital. A high-bandwidth local bus should be used to connect these I/O modules to processing components. The GPU card can be considered as an optional processing component as PPNs can either be implemented only with one or more multi/many-core processors (MCP). The GPU accelerator card is suitable for computationally intensive and memory bound applications. Such applications should also exhibit fine-grain parallelism in order to exploit GPU computational resources. Other applications can profit from sequential and parallel computing features of multi/many-core processors that nowadays integrate also vector processing units for single instruction multiple data (SIMD) execution. Multiple MCPs can be used. However, it is better to use processors with integrated memory controllers that enable non-uniform memory access (NUMA) multiprocessor designs. Such designs tend to improve application scalability in distributed shared memory systems by improving memory hierarchy performance.

With a similar design to PPNs, the RHMU must integrate a radiation hardened processor or FPGA [2, 15]. Memory modules have to be configured in a triple modular redundancy (TMR) fashion to increase data integrity. Additional reliability techniques should be implemented on software. One of the best ways to achieve this is to integrate a fault-tolerant middleware between operating system and software application [16] and to provide redundancy at the execution process or thread level [7, 12].

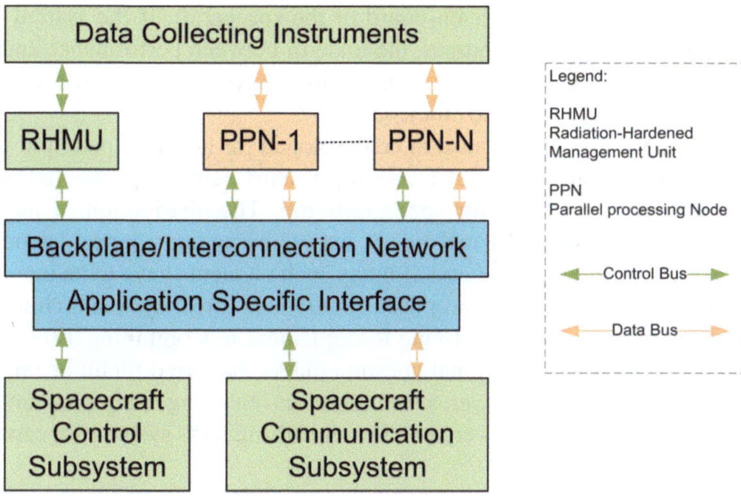

Fig. 1 HPC architecture for future space applications

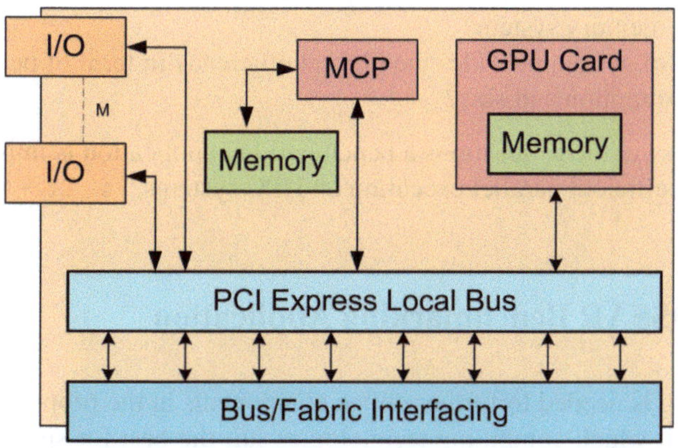

Fig. 2 The parallel processing node

4 Expected Problems in the Proposed Architecture

The proposal in this paper combines various technologies to provide a solution for future space applications. Nevertheless, it is not possible to address all probable issues in the same solution. Being very flexible in design, the architecture can be configured according to various requirement levels in terms of performance and reliability. Main problems that might arise are related to power consumption, dissipated heat, and size.

The integration of many high performance computing components in the same system increases power consumption. Other solutions should be found on how to obtain that amount of power on-board of the spacecraft. If the system is used efficiently without wasting resources, the trade of between performance and power consumption might be acceptable. Applications have to be tailored to specific architectural features to increase efficiency.

Probably new technologies have to be developed for heat dissipation in the lack of air. Emerging hot and cold water cooling systems being implemented on Earth might have to be considered for spacecrafts too. The distribution of processing elements helps in reducing thermal issues, but increases interconnect latencies and complicates communication patterns if many such elements have to be integrated.

It might be hard to apply all proposed solutions in a real system architecture as new problems might arise when trying to implement it. Combining different technologies might add some additional question marks that are difficult or impossible to predict at the moment. However, a few questions regarding the implementation of the cluster of PPNs can be answered by benchmarking HPC systems on earth. Such questions are:

1. How many PPNs should be used?
2. Do we need to integrate accelerators on each PPN?
3. Should the cluster of PPNs be implemented as a shared memory or as a distributed memory system?
4. What kind of system provides the highest efficiency in term of performance per power consumption and size?

In order to answer those questions a benchmarking application is implemented and optimized for efficient parallel execution on HPC systems.

5 The 2DSSAR Benchmarking Application

Benchmarking is needed to further define components in the proposed architecture and to find out which computing technologies suit the best for space applications. Being focused on SAR processing, this section describes the application that is used to benchmark shared memory, distributed memory, and heterogeneous CPU/GPU platforms. The 2-dimensional spotlight SAR (2DSSAR) application is implemented based on formulations given in [17] and uses the structure applied in SSCA benchmarks [1]. It is composed of two stages called synthetic data generator (SDG) and SAR sensor processing (SSP). SDG is used to generate synthetic SAR returns from a uniform grid of point reflectors, which are similar to what would be obtained from a real SAR system. SSP retrieves generated data and reconstructs the resulting SAR image using a spotlight SAR reconstruction method known as spatial frequency interpolation (SFI). This method involves FFT transforms (22 %), compression and decompression loops (7 %), transposition and FFT-shift operations (2 %), and a 2-dimensional interpolation loop (69 %) that converts data from polar

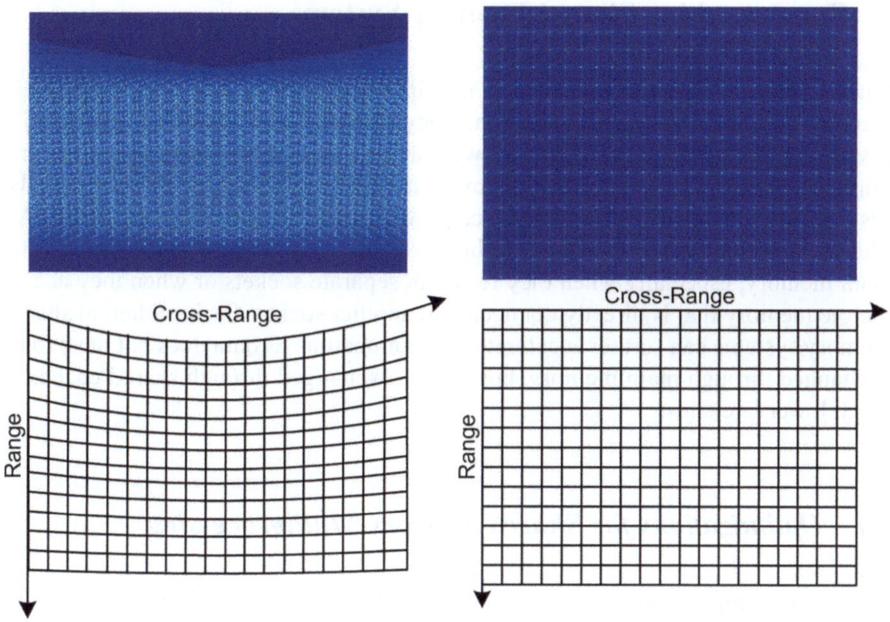

Fig. 3 Image reconstruction from SAR returns

Table 1 SAR sensor data and reconstructed image sizes in each dimension

	SAR sensor data		Reconstructed image	
SCALE	Range	Cross-range	Range	Cross-range
10	3,290	1,600	3,808	2,474
30	9,630	4,800	11,422	7,380
60	19,140	9,600	22,844	14,738

to rectangular coordinates. Figure 3 illustrates visual transformations applied by the reconstruction method. Synthetically generated SDG SAR returns are shown on the left hand side in polar coordinates, whereas the SSP reconstructed image is shown on the right hand side in rectangular coordinates. Swath size in the direction the radar is looking into is represented by the range dimension. Cross-range is the dimension perpendicular to range in flight direction.

2DSSAR can be easily configured to reconstruct images of different sizes, from small to large scale ones. Table 1 shows SAR sensor data and reconstructed image sizes for small, medium, and large scale. We are mainly focused in benchmarking with large scale image reconstruction. However, other scales are used for comparison reasons.

6 Benchmarking Shared Memory Systems

Shared memory systems consist of multiple processor cores that share main memory and use it for communication. They are usually implemented either with a single globally shared memory or with a physically distributed one. The first implementation provides uniform memory access (UMA) for all the cores in the system, whereas in the second one access to memory is non-uniform (NUMA). This means that different cores might have different latencies when accessing main memory, especially when they reside in separate sockets or when they access remote memory that is directly connected to another socket. Cache coherent shared memory systems can further accelerate communication, as data does not have to be exchanged through main memory, but can be exchanged through shared caches at much lower latencies.

6.1 Optimizations for Shared Memory Multiprocessing

The basic computation construct in shared memory systems is the thread. All threads in each process share the same address space and communicate with each other by writing and reading shared variables to and from this shared memory region. OpenMP is used to create, manage, synchronize, and destroy threads in 2DSSAR. For most processing steps in SDG and SSP a parallel region is created to distribute work (loop iterations) over multiple threads. As soon as a parallel region is entered, new threads are forked and joined again in the end of the region. To avoid memory bottlenecks in UMA systems, cache *pre-fetching* is used to hide memory latencies for application parts that have regular access patterns. For large scale image reconstruction the complete data set does not fit in cache. *Blocking* technique is used to divide data into smaller parts that fit in cache. Memory accesses in NUMA systems are scattered to avoid memory latencies. Furthermore, data partitioning schemes that make sure that most frequent accesses take place in local memory are used to reduce remote memory accesses. Main techniques used in these schemes are *thread pinning* and *first touch policy*. Thread pinning technique binds threads to specific cores in order to reduce possible thread migrations that might bring additional remote memory accesses. First touch policy assures that each thread initializing a data object gets the page associated with that data item in the local memory of the processor it is executing on. Hybrid programming is also considered to help in keeping most frequently accessed data in local memory. This is achieved by combining MPI and OpenMP to assure that each multi-core processor accesses local memory the whole time. Threads in each core share only local memory and not memory that is associated with another multi-core processor on another socket of the system.

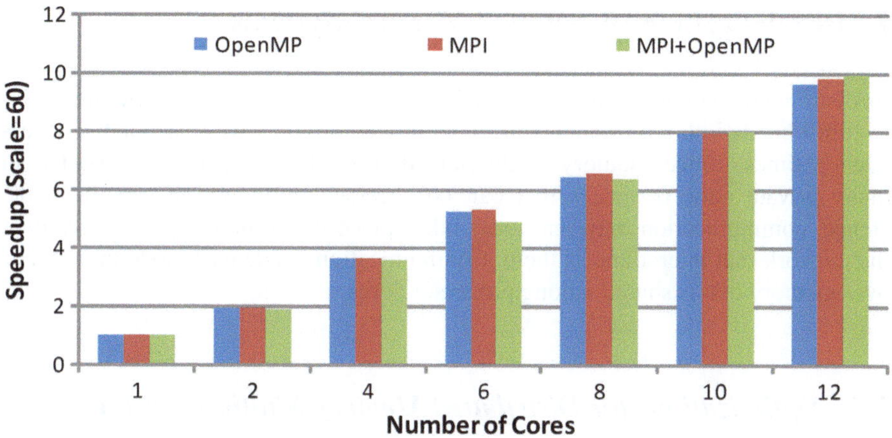

Fig. 4 SAR image reconstruction on ccNUMA platform

6.2 Obtained Results

The platform used for benchmarking purposes is a dual-socket cache coherent NUMA system integrating two Intel Nehalem chip-multi processors operating at 2.93 GHz. Each processor has 6 cores that access directly 18 GB of main memory through an integrated memory controller. This results in a total of 36 GB shared memory. This is a very important feature for 2DSSAR that uses large data sets when reconstructing large scale images. This platform uses special purpose hardware and cache coherency protocols to provide cache coherency across shared memory.

The optimizations discussed in Sect. 6.1 are applied to 2DSSAR, which is first multithreaded using OpenMP pragmas and then parallelized using message passing paradigm with MPI. As shown in Fig. 4, 2DSSAR performs a bit faster when using MPI because distributed memory programming reduces remote memory accesses by having each process sending and receiving only the required amount of data. The only drawback of distributed memory programming is that it requires a lot of programming efforts to achieve desired level of scalability. Hybrid programming combines the efficiency of MPI with the flexibility of OpenMP. MPI is used in 2DSSAR to create and synchronize processes for each socket in the platform (two in this case) and OpenMP is used to manage a number of threads equal to the number of cores in each socket (six in this case). This is much easier to be programmed and it provides comparable performance to other approaches. Indeed, results show that the hybrid version of 2DSSAR performs the best when using all 12 cores in the platform. Results also show that this application can be easily programmed and scaled for shared memory multiprocessing.

7 Benchmarking Distributed Memory Systems

In distributed memory systems, memory is distributed among computing nodes to fulfill bandwidth requirements of many processor cores with acceptable access latency times. Since memory is distributed, the address space is separated in many private address spaces that can be accessed only by one specific node. Hence, communication between nodes takes place by exchanging messages over the network that interconnects them. Communication inside each node takes place over memory that is shared among processor cores.

7.1 Optimizations for Distributed Memory Multiprocessing

The MPI version of 2DSSAR used to benchmark the ccNUMA platform is further optimized for execution on distributed memory systems. The 2-dimensional SAR data set is initially distributed along various processes by chunks of rows. Then, some processing steps of 2DSSAR are optimized so that they can be executed in processes without having to exchange data. This allows concurrent execution of those steps. Nevertheless, there are also some steps that require inter-process communication. Such steps involve FFT-shift, transpose, and reduction operations.

The FFT-shift operation swaps first and third quadrants, but also second and fourth quadrants of the 2-dimensional data. In a peer-to-peer communication pattern, processes that have data from first and second quadrants exchange data with processes that have data from third and fourth quadrants. This communication pattern is implemented using non-blocking send and receive MPI functions that enable overlapping of communications and computations among various processes. The only drawback of such an approach is that additional buffering memory is needed to store data being exchanged. Figure 5 depicts the new communication pattern of the distributed FFT-shift.

Communications are overlapped with computations in transpose operations too. In this case an all-to-all communication pattern and a horizontal and vertical data tiling is needed. Diagonal tiles do not have to be moved, they are simply transposed by the respective process. Other tiles are sent, received, and buffered in a non-blocking way and then finally transposed by the destination process. Figure 6 illustrates the overlapping of the transposition operation with the communication. As shown in Fig. 7, the communication pattern is now an all-to-all one.

After the interpolation loop, overlapped regions between neighbor chunks of rows have to be reduced. A collective reduction operation is very expensive in this case because the amount of overlapped regions is quite small compared to the complete data. A local reduction between neighbor processes is implemented to reduce only overlapped rows and not the whole 2-dimensional data. The reduction is scheduled in an ordered way so that the first process sends data to the second

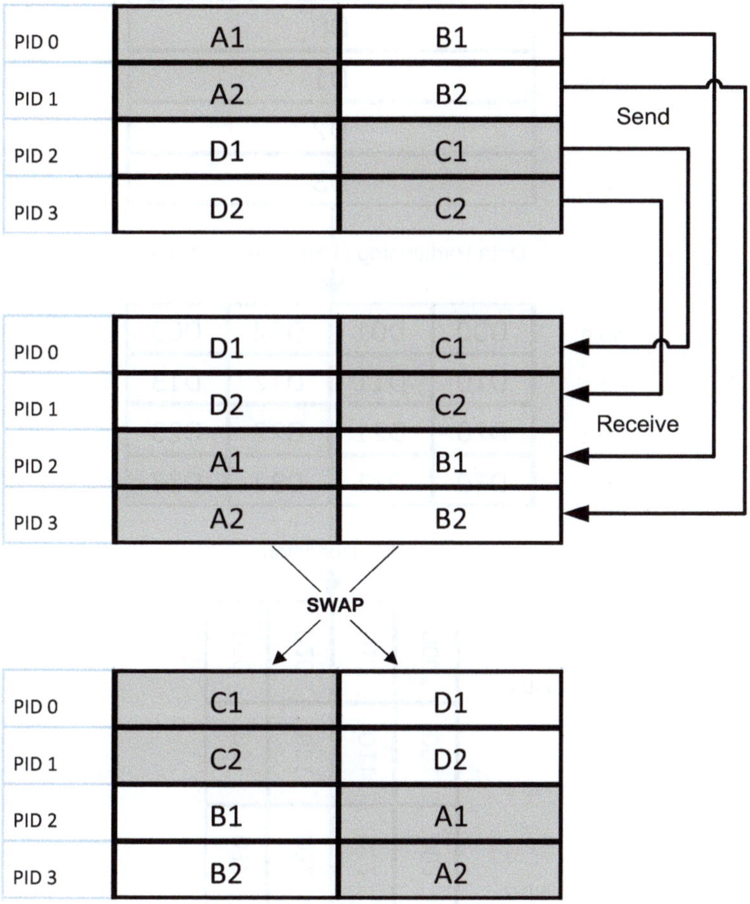

Fig. 5 Inter-process communication and swapping in the FFT-shift operation

process, which accumulates new values with old ones and sends results back to the first process (Fig. 8).

7.2 Obtained Results

Benchmarking results are obtained on the Nehalem Cluster at HLRS Stuttgart. This cluster is composed of 700 NEC HPC-144 Rb-1 Server compute nodes that are interconnected via 24 leaf and 6 backbone voltaire grid director (VGD 4036) switches in a double data rate Infiniband network. In this fat-tree topology, each leaf switch interconnects 30 nodes with each backbone switch through bidirectional links. Each switch provides 36 quad-data rate ports with 40 GB/s throughput.

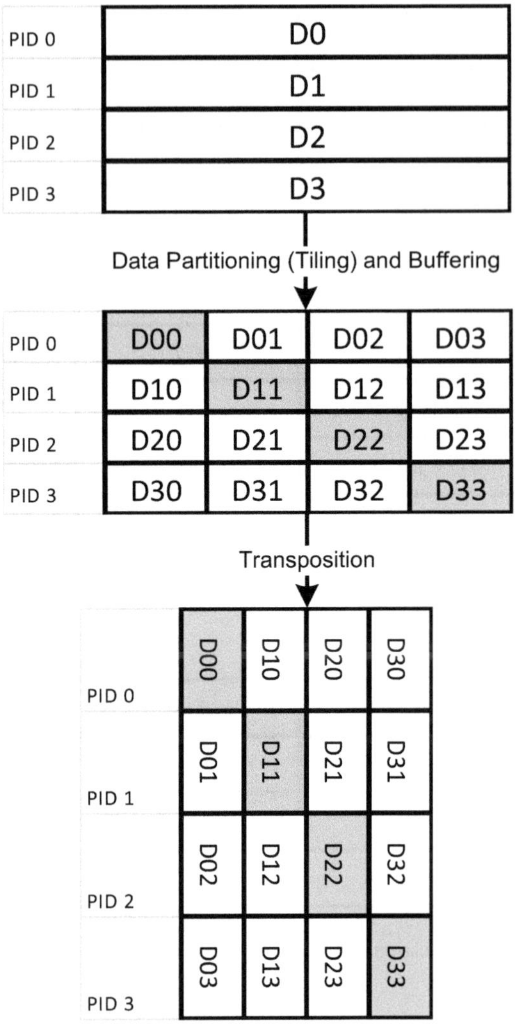

Fig. 6 Overlapping communications and tile-transpositions

Each compute node is a ccNUMA system composed of two quad-core Intel Xeon X5560 CPUs that support 8 threads in simultaneous multithreading and have an 8 MB L3 shared cache. The integrated memory controller communicates with memory at 1,333 MHz offering an aggregate bandwidth of 32 GB/s. Each core normally operates at 2.8 GHz, but in Turbo mode it goes up to 3.2 GHz.

The optimizations discussed in Sect. 7.1 are implemented in all three versions of 2DSSAR that are used to benchmark distributed memory systems. Obtained results are shown for each of them in Fig. 9. With the pure MPI version of 2DSSAR the

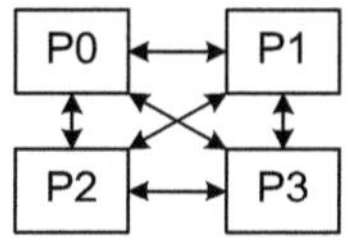

Fig. 7 Inter-process communication in transpose operations

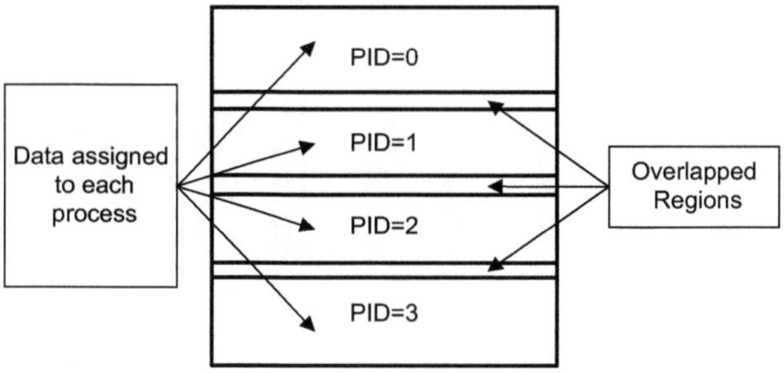

Fig. 8 Overlapped regions

most efficient speedup factor of 23.7 is obtained when using 8 cluster nodes. With this speedup factor that represents 37 % of the theoretical linear speedup the time needed for large scale image reconstruction is reduced from 10 min and 39 to 26 s. This time is reduced to 20 s when using 16 nodes, but the speedup factor of 32 now represents only 25 % of the theoretical linear speedup. This shows that even though higher speedup values are obtained, the scalability of the application when using more than eight nodes is getting lower. The reason for this is the amount of inter-process communications that keeps rising as the number of processes increases.

Unfortunately this is also true for the Hybrid (MPI + OpenMP) version. Even though it performs a little better than the MPI version, the application scalability is not at the desired levels. Elapsed time on 8 nodes is reduced to 21.5 s giving the most efficient speedup factor of 27, which represents 42 % of the theoretical linear speedup. On 16 nodes, elapsed time is further reduced to 18 s, but the speedup factor of 32.3 represents only 25.2 % of the theoretical linear speedup.

In order to use the distributed memory Nehalem Cluster in a more efficient way, a pipelined version of 2DSSAR is implemented. This pipelined version exploits a specific feature of the iterative image reconstruction scenario, in which multiple images get processed in the long run. It is named "Pipelining" because different image reconstructions are pipelined along different cluster nodes. This technique is suitable to be used in distributed memory environments with multiple nodes that have enough memory resources to accommodate the complete required data set. Multiple processes communicating via message passing are distributed along different nodes, but the parallel execution inside each node is selected to be done by OpenMP threads, since the hybrid implementation showed a slightly better parallel performance than the pure MPI one.

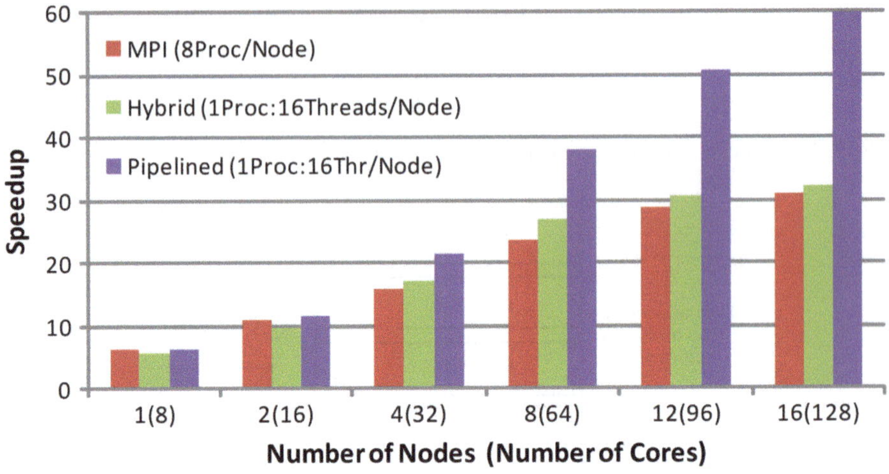

Fig. 9 SAR image reconstruction on Nehalem Cluster HLRS

By having each node processing a single image, the time spent for inter-node communication is reduced. It might take some time to output the first image in pipelined 2DSSAR, but then other images are generated much faster. The number of images processed in one round depends on the number of nodes used in the cluster. As shown in Fig. 9, the profit of using pipelining starts to become visible when using more than four nodes. As the number of images that have to be reconstructed increases, the number of processes and the number of nodes being used increases too. More important is that obtained parallel performance increases too. For the pipelined 2DSSAR, the most efficient speedup factor is also the highest one, which is obtained when using 16 nodes. To summarize, a speedup factor of 60 is obtained by using 128 cores in 16 nodes of Nehalem Cluster and the average elapsed time per image reconstruction is reduced to 10.7 s. This means that the distributed memory system is used with an efficiency of 47 %.

8 Benchmarking Heterogeneous CPU/GPU Systems

Heterogeneous CPU/GPU systems are taken into consideration because GPUs offer higher computational power and bandwidth capacity for specific applications that can profit from their architectural features. Performance of data-intensive applications never scales linearly on many-core CPUs due to constraints in memory bandwidth. On the other side, GPUs are considered data throughput processors with much more favorable computational power to bandwidth ratio. GPUs can deliver higher computational power for parallel workloads. One of the main drawbacks of heterogeneous CPU/GPU computing is the throughput bottleneck in the PCI

Express (PCIe) interconnection between host (CPU) and device (GPU). Frequent and unnecessary data movements between host and devices should be minimal, but if data does not fit into GPU memory such movements cannot be avoided. It is sometimes necessary to create optimal communication patterns that overlap with computations on both CPU and GPU.

The most interesting feature that makes heterogeneous systems suitable for high performance computing is that it combines two different computing technologies that differ in design philosophy. CPUs integrate sophisticated control logic and large cache memories to increase instructions throughput and hide memory access latencies. On the other side, GPUs are designed to have more chip area for floating-point calculations and very high memory bandwidth. Therefore, CPUs are suitable for optimized sequential and coarse grained parallel code, whereas GPUs for fine grained parallel code.

8.1 Optimizations for Heterogeneous CPU/GPU Processing

Heterogeneous computing is very beneficial to SAR processing and especially to 2DSSAR. However, some additional optimizations are needed to efficiently use these features to improve application performance. A deep understanding of CPU and GPU computing is needed. Moreover, the system layout should be analyzed as a first step towards efficient resource usage. Recent heterogeneous systems combine in a single platform one ccNUMA CPU subsystem and one GPU accelerator module. In general, the number of CPUs in the ccNUMA system is equal to the number of GPUs in the accelerator module, so that each CPU is interfaced directly to one GPU. The detailed architecture of the benchmarked platform in shown in Fig. 10. The ccNUMA subsystem integrates two quad-core Intel Nehalem CPUs running at 2.13 GHz. Each CPU has direct access to 6 GB of memory, building 12 GB of main memory, which is shared by 8 cores in the ccNUMA subsystem. The accelerator module is composed of 2 GPU boards, each containing an NVIDIA Tesla GPU, which accesses 6 GB of GPU global memory with a memory bandwidth of 144 GB/s (4.5 times higher than CPU memory bandwidth). Most limiting parameter in the whole platform is the GPU PCIe bandwidth that can reach only up to 8 GB/s.

Optimizations are initially applied to 2DSSAR to improve communication and memory access patterns in the heterogeneous platform. Remote memory accesses should be reduced in the ccNUMA subsystem and transfers between GPU memory and remote CPU memory (red line) should be avoided because they take longer than local transfers (green line). This problem can be partially solved by pinning threads that manage GPU contexts on different CPUs, so that they can use separate CPU memories. If only one GPU is used, remote accesses might occur only when CPU threads initialize data on remote memory and same data has to be copied to GPU memory. If both GPUs are used, additional remote accesses might occur when data has to be copied from one device to the other. Furthermore, the limiting PCIe bandwidth becomes a bottleneck especially when the data set does not fit

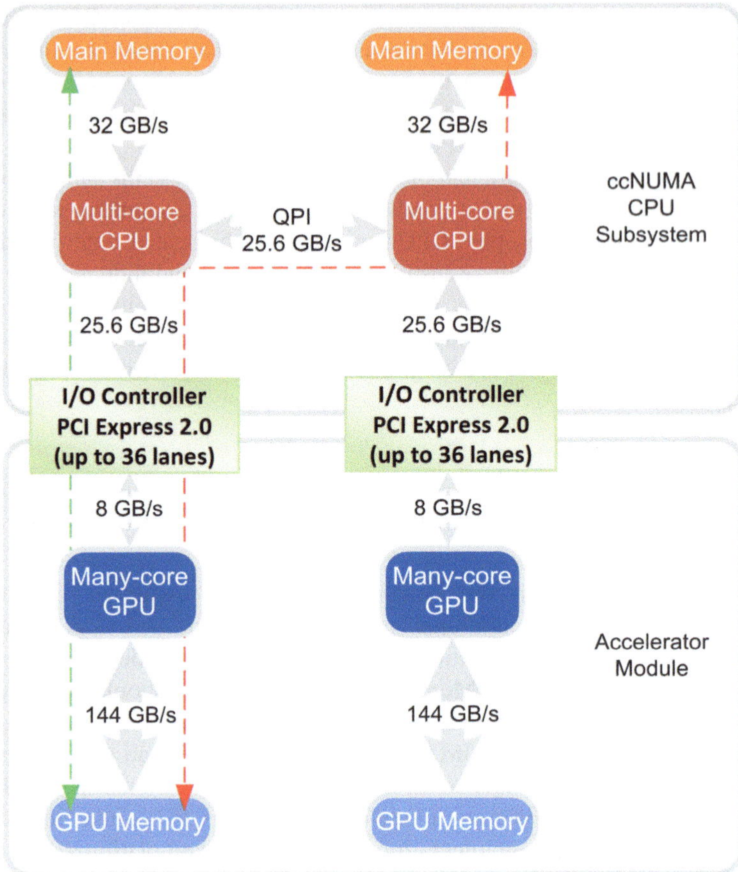

Fig. 10 Heterogeneous CPU/GPU architecture

into GPU memory. That is actually the case for large scale image reconstruction in 2DSSAR. Specific communication patterns have to be implemented to avoid additional latencies in this case.

In order to be able to run 2DSSAR on the accelerator module, it is first ported to CUDA C programming language. A 2-dimensional tiling technique is used for workload distribution. Each tile is addressed by block-id and each thread in each tile is addresses by thread-id. Indeed, this tiling technique is easily implemented for most processing steps of 2DSSAR. Other steps that involve FFT transforms make use of cuFFT [8] library functions. cuFFT implements batch execution for doing multiple one-dimensional transforms in parallel and hides underlying details of implementation to the programmer.

CUDA C lacks some operations on complex data that we have implemented using basic operations on float data. Some specific atomic operations on complex data are implemented by using atomic compare and swap operations on float data, which means that a preliminary extraction of real and imaginary parts of the complex value is needed. Atomic operations are used to avoid race conditions in the interpolation loop. They are mostly used to prevent race conditions between different threads within the same thread block in shared memory and just in a few cases in global memory. Performance is not impacted so much because shared memory is quite fast and private to each block. This means that atomic operations on different blocks do not interfere with each other and are executed in parallel.

Transcendental instructions such as *sine*, *cosine*, and *square root* are used for better performance since they execute on Special Function Units (SFUs). Each streaming multiprocessor contains four SFUs with separated pipeline from dispatch unit. This improves performance since dispatch unit issues instructions to other execution units while SFUs are occupied.

8.2 Obtained Results

Different incremental versions are implemented, beginning with small and large scale image reconstruction, going to a version that uses both CPU and GPU for SAR processing, and concluding with the version that uses multiple GPU devices. For small scale image reconstruction, data is copied initially to GPU memory and copied back to main memory after it has finished being processed by the GPU device. Since the data set fits into GPU memory all processing steps are executed consecutively with no communications in between. However, for large scale image reconstruction the data set does not fit into GPU memory, especially in the interpolation loop. The iterations of this loop are divided in three parts that are executed as separate kernels on the GPU. Each kernel now needs only one third of data on GPU memory. Only the first data movement from the CPU to the GPU is expensive. Other movements do not impact performance since they are overlapped with computations.

Figure 11 shows results obtained on the benchmarked heterogeneous platform. Speedups are calculated by referencing the best sequential execution on a single CPU core. The best sequential code is also vectorized for optimal execution on CPU SIMD units. First group of results shows the CPU Sequential reference speedup. Different image scales are analyzed for each version of 2DSSAR. Second group (column) shows speedups obtained on the ccNUMA subsystem with eight threads on eight cores. These speedup values are lower than the ones shown in Fig. 4 due to the impact of vectorization and other sequential optimizations on the parallelism of the application. Third group shows speedups obtained with 16 threads in simultaneous multithreading, which improves performance just a bit. Memory bandwidth constraints in the ccNUMA subsystem give the difference in obtained speedups for small scale and large scale image reconstruction.

Results on the complete heterogeneous platform start from the fourth group (marked *GPU*) that shows speedups obtained when using only one GPU for SAR processing. In this case the CPU is used only for managerial purposes. There is again a higher speedup for small scales, but now the difference is quite small. All results on the GPU show higher speedups than the CPU ones. They become even higher

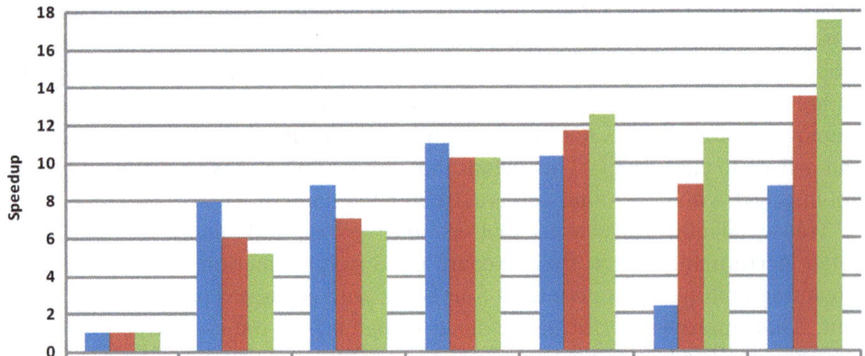

	CPU Sequential	CPU 8 Threads	CPU 16 Threads (SMT)	GPU	CPU + GPU	2 GPUs	2 GPUs Pipelined
Scale=10	1	7.95	8.82	11.05	10.37	2.35	8.71
Scale=30	1	6.04	7.01	10.29	11.70	8.84	13.44
Scale=60	1	5.21	6.37	10.24	12.53	11.30	17.49

Fig. 11 SAR image reconstruction on the heterogeneous CPU/GPU platform

when using also the CPU for SAR processing (group *GPU + CPU*). However, the resulting speedup values are not near the sum of individual CPU and GPU speedups due to communications needed between host and device.

Last two groups of results show speedups obtained when using both GPUs on the platform. The group of results marked *2 GPUs* depicts speedups obtained when distributing the workload of a single image reconstruction on two devices. Due to high communication overheads 2DSSAR does not scale very well in this case. In some processing steps, it is necessary to exchange data between GPU memories. Such exchanges give additional overhead as they take place through main memory. This overhead impacts performance quite a lot, especially for small scale image reconstruction that does not present too much computations. The group marked *2 GPUs Pipelined* depicts speedups obtained when reconstructing one image in each device. The latter case offers the best performance of all, especially for large scale image reconstruction. Better performance is obtained because there is no data exchange between devices and each device processes independently a separate image. Each of them receives SAR sensor data from the CPU and returns a separate reconstructed SAR image.

9 Conclusions

Obtained benchmarking results form a helping asset for spacecraft computing platform designers, especially for those who deal with designs of future space applications that will need high performance computing for on-board processing. Each benchmarked platform gives an improved parallel performance within expected values. Small and mid-sized shared memory platforms are sometimes insufficient to provide the required performance level. Large scale distributed memory platforms can be considered as an alternative solution for the required performance, but the efficiency rate in terms of power consumption, size and heat dissipation goes down. Moreover, additional efforts are needed to parallelize the application so that expected performance objectives can be achieved. Last but not least an important alternative is the heterogeneous CPU/GPU platform that offers much better performance than shared memory one without wasting resources and compromising system efficiency. The final recommendation for space applications with similar processing steps to 2DSSAR would be to use heterogeneous CPU/GPU systems for on-board processing. Future CPUs with wide SIMD units or even integrated GPUs would also be interesting for such applications.

References

1. Bader, D.A., Madduri, K., Gilbert, J.R., Shah, V., Kepner, J., Meuse, T., Krishnamurthy, A.: Designing scalable synthetic compact applications for benchmarking high productivity computing systems. Cyberinfrastruct. Technol. Watch (2006). http://gauss.cs.ucsb.edu/publication/ctwatch-ssca.pdf
2. Berger, R.W., Bayles, D., Brown, R., Doyle, S., Kazemzadeh, A., Knowles, K., Moser, D., Rodgers, J., Saari, B., Stanley, D., Grant, B.: The RAD750TM-a radiation hardened PowerPC processor for high performance spaceborne applications. In: 2001 IEEE Aerospace Conference, Big Sky, vol. 5, pp. 2263–2272 (2001). doi:10.1109/AERO.2001.931184
3. Choi, I., Zhao, M., Yang, X., Yeung, D.: Experience with improving distributed shared cache performance on Tilera's tile processor. Comput. Archit. Lett. **10**(2), 45–48 (2011). doi:10.1109/L-CA.2011.18
4. Crago, S.P., Kang, D.I., Kang, M., Kost, R., Singh, K., Suh, J., Walters, J.P.: Programming models and development software for a space-based many-core processor. In: 2011 IEEE Fourth International Conference on Space Mission Challenges for Information Technology (SMC-IT), Palo Alto, pp. 95–102 (2011). doi:10.1109/SMC-IT.2011.29
5. Garland, M., Kirk, D.B.: Understanding throughput-oriented architectures. ACM Commun. **53**, 58–66 (2010)
6. Ginosar, R.: The plural architecture-shared memory many-core with hardware scheduling (2012). http://webee.technion.ac.il/~ran/papers/PluralArchitectureJan2012.pdf
7. Gold, B.T., Falsafi, B., Hoe, J.C., Mai, K.: REDAC: distributed, asynchronous redundancy in shared memory servers. Technical report, Computer Architecture Lab at Carnegie Mellon (CALCM) (2008)
8. Govindaraju, N.K., Lloyd, B., Dotsenko, Y., Smith, B., Manferdelli, J.: High performance discrete fourier transforms on graphics processors. In: Proceedings of the 2008 ACM/IEEE Conference on Supercomputing, SC '08, Austin, pp. 2:1–2:12. IEEE Press, Piscataway (2008). http://dl.acm.org/citation.cfm?id=1413370.1413373
9. Harzallah, K., Sevcik, K.C.: Hot spot analysis in large scale shared memory multiprocessors. In: Proceedings of the Supercomputing Conference, Portland, pp. 895–905 (1993). doi:10.1109/SUPERC.1993.1263548
10. Intel: Intel's teraflops research chip. Website. http://download.intel.com/pressroom/kits/Teraflops/Teraflops_Research_Chip_Overview.pdf

11. MPI-Forum: The message passing interface (MPI) standard. Website. http://www.mpi-forum.org/. Last accessed in June 2012
12. Mukherjee, S.S., Kontz, M., Reinhardt, S.K.: Detailed design and evaluation of redundant multi-threading alternatives. In: Proceedings 29th Annual International Symposium on Computer Architecture, Anchorage, pp. 99–110 (2002). doi:10.1109/ISCA.2002.1003566
13. NVIDIA: Cuda parallel programming made easy. Website (2012). http://www.nvidia.com/object/cuda_home_new.html
14. OpenMP-ARB: The OpenMP® API specification for parallel programming (2011). http://openmp.org/wp/openmp-specifications/
15. Pham, C., Malcom, H., Maurer, R., Roth, D., Strohbehn, K.: LEON3FT Proton SEE test results for the solar probe plus program. In: IEEE Radiation Effects Data Workshop, Las Vegas, pp. 1–4 (2011). doi:10.1109/REDW.2010.6062535
16. Ramos, J., Samson, J., Lupia, D., Troxel, I., Subramaniyan, R., Jacobs, A., Greco, J., Cieslewski, G., Curreri, J., Fischer, M., Grobelny, E., George, A., Aggarwal, V., Patel, M., Some, R.: High-performance, dependable multiprocessor. In: 2006 IEEE Aerospace Conference, Big Sky, p. 13 (2006). doi:10.1109/AERO.2006.1655959
17. Soumekh, S.: Synthetic Aperture Radar Signal Processing with MATLAB Algorithms. Wiley, New York (1999)
18. Suess, M., Grafmueller, B., Zahn, R.: A novel high resolution, wide swath SAR system. In: 2001 IEEE International Geoscience and Remote Sensing Symposium, Sydney, vol. 3, pp. 1013–1015 (2001). doi:10.1109/IGARSS.2001.976731
19. Villalpando, C., Rennels, D., Some, R., Cabanas-Holmen, M.: Reliable multicore processors for NASA space missions. In: 2011 IEEE Aerospace Conference, Big Sky, pp. 1–12 (2011). doi:10.1109/AERO.2011.5747447
20. Walters, J.P., Kost, R., Singh, K., Suh, J., Crago, S.P.: Software-based fault tolerance for the Maestro many-core processor. In: 2011 IEEE Aerospace Conference, Big Sky, pp. 1–12 (2011). doi:10.1109/AERO.2011.5747448
21. Wentzlaff, D., Griffin, P., Hoffmann, H., Bao, L., Edwards, B., Ramey, C., Mattina, M., Miao, C.C., Brown, J.F., Agarwal, A.: On-chip interconnection architecture of the tile processor. IEEE Micro **27**(5), 15–31 (2007). doi:10.1109/MM.2007.4378780

Requirements Engineering for Computational Seismology Software

Yang Li, Bernd Bruegge, Simon Stähler, Nitesh Narayan, and Heiner Igel

Abstract Many seismological software applications are developed to support for instance studies of earthquake scenarios, seismic exploration surveys or hazard analysis. Most of these applications were developed in isolation with focus on algorithmic performance, and less emphasized on software's comprehensibility and maintainability. However, requirements engineering practices, which help to describe the applications from a high level point of view, are mostly ignored. As a result, trying to reuse these algorithms in larger systems, developers face difficulties in comprehending, modifying, adapting and integrating the applications. This paper presents our work of introducing the concepts of requirements engineering to seismological research projects. Requirements describe what a software system is expected to do and to be. They are used to communicate between scientists from different domains (e.g. seismology and computer science) to achieve a common understanding for developing the software. Requirements also provide a basis for other development activities, such as software comprehension, software design, implementation and maintenance. We present a model-based requirements engineering approach that incorporates abstractions and notations from the seismology domain. We describe two requirements patterns, which facilitate the identification and specification of requirements in seismological software development. We use a dynamic rupture example to illustrate how seismologists can apply our approach.

Keywords Seismology • Scientific software development • Requirements engineering • Modelling

Y. Li (✉) · B. Bruegge · N. Narayan
Department of Informatics, Technische Universität München, Munich, Germany
e-mail: liya@in.tum.de; bruegge@in.tum.de; narayan@in.tum.de

S. Stähler · H. Igel
Geophysics, Department of Earth and Environmental Science, Ludwig-Maximilian-University, Munich, Germany
e-mail: igel@geophysik.uni-muenchen.de; staehler@geophysik.uni-muenchen.de

M. Bader et al. (eds.), *Advanced Computing*, Lecture Notes in Computational Science and Engineering 93, DOI 10.1007/978-3-642-38762-3_8,

1 Introduction

Software applications are developed for different areas of computational seismology such as earthquake scenarios, volcanic processes, or seismic exploration surveys. Moreover, they are widely developed and used for hazard analysis and risk management. High performance computing (HPC) is also incorporated into many seismological applications, to efficiently handle huge data volumes and intensive computations.

In the past, such software has usually been developed ad hoc, often in the context of implementing novel numerical methods and improving their performance. Since the focus is on answering scientific or numerical questions, little effort is put into software's comprehensibility, modularity, or changeability. For example, software does not have clear or complete documentation. It is difficult to comprehend the software [14]. Moreover, the functional modules are highly coupled and modules cannot be understood easily in isolation. As a result, changes or extensions to functionality are difficult to perform. Finally, source code has been recycled (cloned, modified and reused) repeatedly over decades, often exceeding their original scope and potentially leading to a plethora of error-prone code. This software development style has served the seismological community for many years. However, with the increasing sophistication of programming languages, algorithmic libraries and high-performance computing, the quality demands on software are increasing tremendously. On the other hand, large-scale interdisciplinary collaborations between seismological software projects and high performance computing projects, along with other scientific projects, emerge rapidly. This requires well-established common understanding of the software to better support collaborative development.

Seismological software has many requirements such as a particular numerical method to be developed and required precision of calculation. The requirements also cover expected functionalities of computation control and user interaction. We claim that these requirements need to be clearly elicited and well managed. Requirements correspond to sophisticated code representing physical theories. Also they provide a baseline for designing, implementing, validating and maintaining software. Clearly defined requirements are analyzed to help decompose the system based on functionalities and choose suitable APIs and platforms. Furthermore, specified requirements refine the ideas from project partners and need to be agreed upon.

However, during seismological software development, traditional requirements engineering practices are mostly ignored. Requirements are often specified implicitly or unclearly without taking much account of future usage. Due to lack of time and good understanding of requirements engineering concepts [8], state-of-the-art requirements engineering techniques and tools are difficult to be applied without proper adjustments [15]. Methodologies are required to provide the flexibility to document requirements with the possibility of refinement and extensibility [1]. Documented requirements are also needed to cope with requirements evolution over time and traceability links between requirements and other artifacts involved in software development, such as source code and test cases.

To meet these needs, we introduce a model-based requirements engineering approach targeted at the computational seismology domain. This work is built on our existing research work about requirements engineering for scientific computing projects [15, 16]. The approach is based on a meta-model that defines domain-specific abstractions, which are systematically extracted by semi-automatically analyzing literature from the seismology domain. To further facilitate capturing requirements productively, we also identify requirements patterns that reuse existing knowledge of seismology software.

The methodology presents two main advantages for seismologists. First, it can be easily understood and applied in computational seismology projects. Second, it draws attention to requirements engineering, which has historically attracted little interest in seismological projects but in fact is crucial to seismology software development. Finally, it is applicable and practical in seismology projects.

The paper is structured as follows: In Sect. 2, we present our model-based approach and describe how to create a requirements meta-model tailored for computational seismology. The meta-model, and two patterns for requirements elicitation and specification are presented in Sect. 3. In Sect. 4 we use a dynamic rupture example illustrate the applicability and practicability of the approach. We briefly compare our approach to other techniques in Sect. 5. Section 6 presents conclusions and future work.

2 Approach

To facilitate requirements engineering for computational seismology, we extend the meta-model proposed in our previous research work of requirements engineering for scientific computing projects. In this section, we present the basic idea of a model-based requirements engineering approach and how the meta-model is extended using domain knowledge.

2.1 Model-Based

A model describes a subject in a simpler and more accessible way to support human understanding and reasoning [10]. It is also used to communicate understanding. In software engineering, we construct models to describe the system under construction. Various modeling languages and notations are used to describe a system from different perspectives, such as describing the functions of the system, the interaction between users and the system and the underlying structure of the system. A meta-model defines the syntax of how to create a model.

In requirements modeling, requirements are expressed in terms of models. A requirements model is at a higher level of abstraction than a textual requirements specification. A model matches more closely the entities, relationships, behavior and constraints of the problem to solve [3]. It can better deal with the high complexity and frequent change in scientific projects. Models can support reusability and extensibility. For example, a project-specific requirements model is created by reusing common parts of an existing model and specializing generic model elements.

We define a meta-model (introduced in [15, 16]) describing artifact types and their relations for modeling requirements in scientific computing projects. This makes modeling less complicated and reduces the learning effort for scientists. It also facilitates the communication across the domain boundary between multi-disciplines, such as the scientific computing domain and the software engineering domain. Aligning to standard modeling languages, it promotes model-driven development in scientific computing projects.

2.2 Domain Knowledge

Domain knowledge plays an important role in eliciting requirements of high quality [12]. We use domain literature and ontology to extend the defined meta-model to a sub-domain of science. In computer science, ontologies are frequently used to describe an application/problem domain. Ontology consists of concepts specific to a particular domain and relations between these concepts. Ontologies are usually organized in taxonomies [7]. They provide a commonly agreed understanding of a domain, which may be reused and shared across applications and groups [2, 6]. Meta-models are closely related to ontologies. Both are often used to describe and analyze the relations between concepts. To reuse the large body of existing domain knowledge presents as text, we perform text analysis on domain literature to extract ontologies, in a semi-automatic fashion utilizing natural language processing. Later the extracted ontology is transformed to a requirements meta-model for a particular sub-domain.

Figure 1 shows the steps of performing text analysis to extend the meta-model for computational seismology.

- To prepare the corpus for text analysis, we select literature of the SPECFEM3D software [4], the VERCE project [21], the SeisSol software [11] and other seismology software projects. In particular, we focus on collecting user manuals and project reports as the input corpus, since they both contain information about the software's functions and properties.
- A top-level ontology is created based on defined classes from the introduced meta-model (model elements) and their associations. The top-level ontology will be populated by analyzing the given corpus of the computational seismology literature.

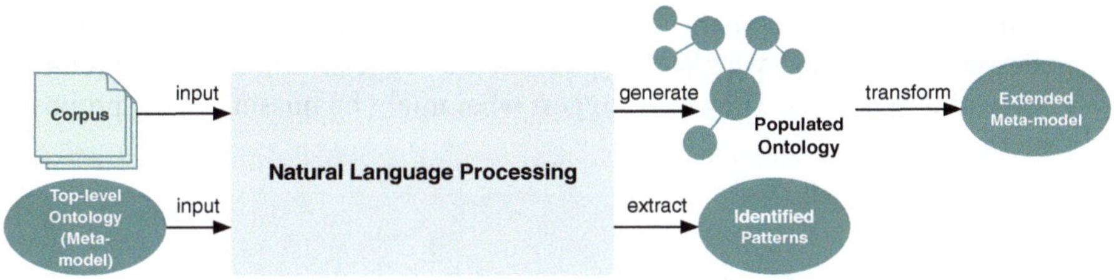

Fig. 1 Steps of text analysis

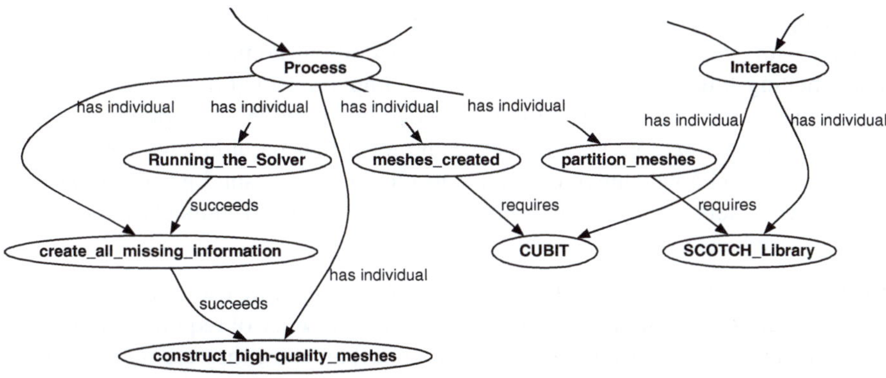

Fig. 2 Visualization of a part of the populated ontology

- We apply natural language processing to extract knowledge from the corpus. The concepts within the corpus are extracted by linguistic pre-processing (tokenization, POS tagging etc.), followed by named entity recognition [18]. Named entity recognition identifies the key terms in the text and relates to the top-level ontology. This is carried out using rule-based grammar or machine learning techniques. For instance, "partition meshes" in the text is an extracted concept, which is a required process of a seismology software system. Relations between concepts are also detected.
- A populated ontology is generated by natural language processing, which consists of extracted concepts from the corpus and detected relations (see Fig. 2). The populated ontology is transformed to an extended meta-model for computational seismology. The individuals in the ontology are generalized at an appropriate level of abstraction and added in the meta-model. Domain experts review the populated ontology and the extended meta-model to assure its correctness.
- Occasionally, similar contexts in the corpus are identified. We summarize information about the concepts and relations in those contexts. We analyze and structure such information into requirements patterns [17]. Therefore, they can be used to perform requirements engineering tasks in similar contexts across seismology projects. The patterns are compliant with the extended meta-model.

We also interview developers of seismology software. They validate the patterns' understandability, and whether the patterns are suggestive in their requirements engineering practices. They also suggest what might be missing in the proposed pattern.

3 Customization for Computational Seismology

We tailor the requirements modeling approach to meet needs in the computational seismology domain. Therefore, requirements engineering can be practiced in the domain with greater cognitive effectiveness. Common domain knowledge stimulates eliciting hidden issues that are actually important to software development. This section presents the resulting extended meta-model by applying natural language processing. We also introduce the identified requirements patterns that capture existing knowledge of requirements eliciting for seismological software.

In order to help the readers to easily understand the (meta-) models and patterns, we first introduce how they are presented.

Model Presentation In the remaining parts of this paper, we use Unified Modeling Language (UML) class diagrams to illustrate the essence of requirements models and patterns, i.e. involved requirements (nodes) and how they are related (lines). In UML, a hollow diamond shape graphically represents a "has a" relationship. A hollow triangle shape represents a generalization relationship ("is a"). Each requirement in the class diagrams can have a textual description attribute to indicate detailed information of each requirement.

Pattern Presentation Each pattern is organized in parts, namely, what is the *name* of the pattern, the pattern can be used in what kind of *activities*, who are the *stakeholders*[1] of the requirements, what is the typical *problem* faced by the stakeholders, what are the *forces* behind the emergency of the pattern, what is the *solution* to solve the mentioned problem and balance the forces, and finally how to *apply* the solution.

3.1 Extended Meta-model

The extended meta-model consists of scientific knowledge modeling and requirements modeling. We detail each part and describe how elements from each part correlate with each other.

[1] A person with an interest or concern in a requirement

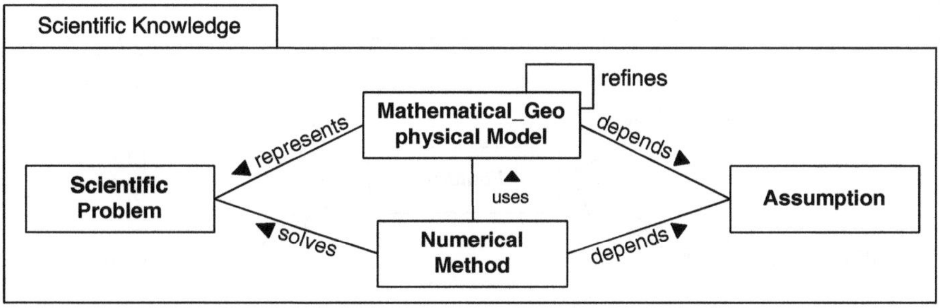

Fig. 3 UML class diagram of scientific knowledge modeling

3.1.1 Scientific Knowledge Modeling

Typically in computational seismology, a seismology-related ScientificProblem is first defined (see Fig. 3). Then Mathematical/GeophysicalModels are created to represent the scientific problem in geophysical models and mathematical terms. In the mathematical modeling process, governing principles and physical laws are determined and factors, which influence the scientific problem, are identified. In order to obtain computational solutions, NumericalMethods for a mathematical/-geophysical model are applied to solve the problem. Mathematical/geophysical models and numerical methods are created on Assumptions.

As an example, we want to solve the scientific problem, simulating seismic wave propagation at the local or regional scale. The problem can be represented by a mathematical/geophysical model, a three-dimensional elastic material volume model. This problem can be solved by the spectral element method (SEM) with explicit time integration.

3.1.2 Requirements Modeling

Figure 4 depicts the requirements modeling part. Features describe desirable properties, qualities or characteristics that are end-user visible and represent an abstract view of the expected solution. We model features to perform a preliminary acquisition step for requirements engineering [9]. For instance, a feature is "a fast and accurate simulation of seismic wave propagation on clusters of GPU graphics cards". A feature can be further refined and be detailed into the realizing Requirements.

An influential factor in scientific computing software development is Hardware. To better utilize the computation power, scientific computing software applications are often hardware dependent. This is especially true for supercomputing applications. Another key element associated with feature is Interface. An interface can either be a software interface or a user interface. The former provides an interface for external libraries and the latter supports end-user interaction with the software

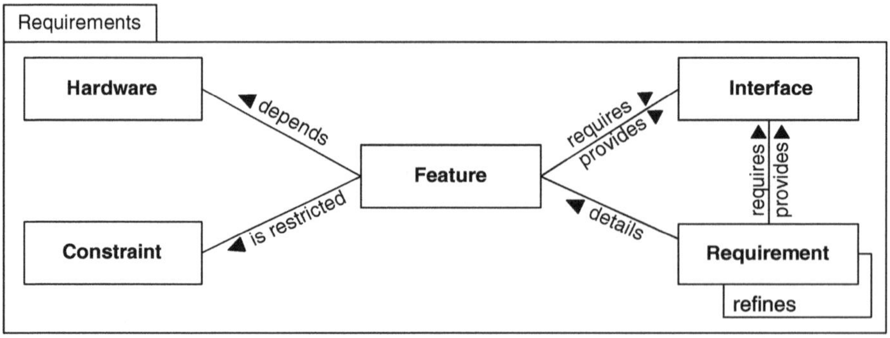

Fig. 4 UML class diagram of requirements modeling

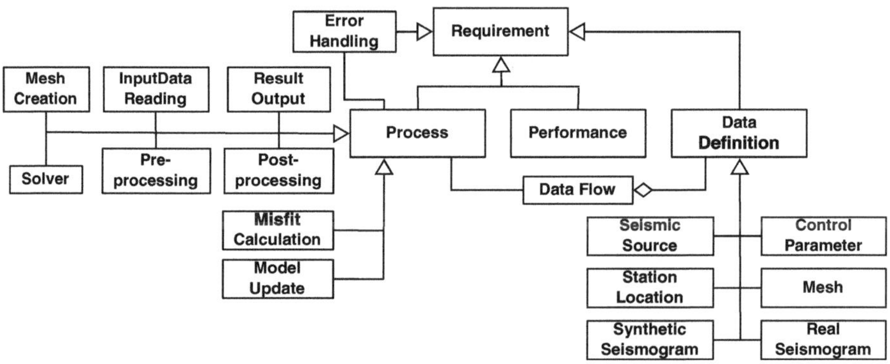

Fig. 5 UML class diagram of types of requirements

itself. Besides the above mentioned limitations, any other Constraints that will limit the developer's design choices, such as the implementation language, have to be identified explicitly as well.

Subclasses of requirement are shown in Fig. 5. Requirement has subclasses of process and performance. Processes specify the functions of processing data, including the algorithm or mathematical expression of each process. A process can be parallel, to perform high performance computing tasks, in contrast to a sequential process. A process can precede or succeed another process. Typical processes involved in computational seismology software are InputDataReading, ResultOutput, MeshCreation, Solver, Misfit Calculation, ModelUpdate, Pre- and Post-processing.

DataDefinition defines the data to be processed. It contains information about for example data type, format, range and accuracy. Data such as SeismicSource, SyntheticSeismogram, StationLocation, Mesh and ControlParameter often need to be defined. This information is used by the process to better manage the data flow. Furthermore, it is also beneficial for data distribution in parallel computing.

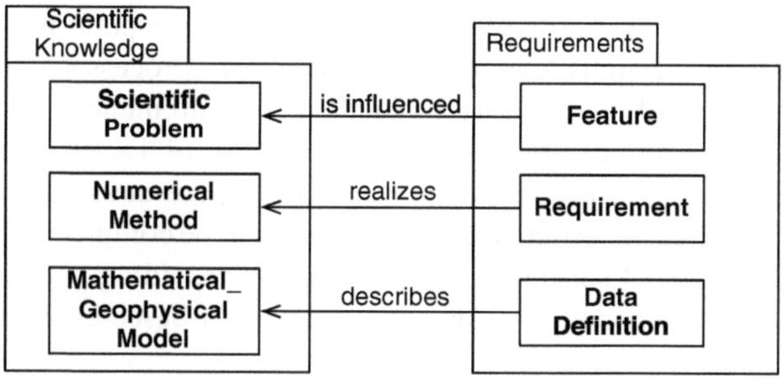

Fig. 6 Links between two parts

There are also subclasses of process to satisfy the needs for scientific computing projects.

Referring back to the example, we can identify the data flow as "input the station location, computational mesh and the parameters describing the seismic source and earth, output the synthetic seismograms". The data definition for a given station location is specified as "station latitude and longitude should be provided in geographical coordinates. The width of the station label should be no more than 32 characters, and the network label should be no more than 8 characters". Other involved data are also specified accordingly. We define a requirement of the type process as "given seismic waves should be simulated numerically by the spectral element method". This requirement can be further refined to steps of the computation process, e.g. mesh creation and solving the wave equation.

3.1.3 Integration

So far, we have already presented the two parts of the proposed model, the scientific knowledge part and the requirements part, respectively. In this section, we discuss how these two parts are integrated. Figure 6 shows links between the two parts. A Feature is influenced by a ScientificProblem. Desired qualities of the problem solving procedure are defined in features. To detail the features, Requirements are created to realize suitable NumericalMethods for the corresponding Mathematical/GeophysicalModel, which is described by DataDefinition. Links within each part and links between the two parts establish the traceability support of this meta-model.

3.2 Identified Patterns

Requirements patterns are used to productively capture desired functionalities and properties of a system by reusing knowledge and can be refined with design and implementation details [13, 22]. We identify requirements patterns targeted at the seismology domain through the mentioned text analysis process and manual review. Existing knowledge of eliciting and managing requirements for seismological software is reused. Patterns stimulate the awareness of issues, which might be hidden but are actually critical to seismological software development. Therefore, requirements can be easily established and they tend to be complete. The patterns are subjected to the needs of seismological software by providing generic abstractions for specialization as per requirements need and pattern-based extension of existing requirements.

In the following, we present two commonly used requirements patterns for seismology software applications, namely, the forward simulation pattern and the data access pattern. It is a revised version of the patterns we described in [17].

3.2.1 Forward Simulation Pattern

Forward simulation (forward modeling) in seismology is a numerical computation process of theoretical or synthetic seismograms, for a given geological model of the subsurface. Forward simulation is frequently applied in computational seismology systems. Although many forward simulation techniques are based on different numerical methods, their structures are similar to organize into a pattern.

Name: Forward Simulation Pattern

Activity: The pattern can be used during requirements elicitation and specification, when forward simulation needs to be developed in a system. For example:

- During requirements elicitation, stakeholders need a starting point to elaborate their requirements.
- Requirements specification of forward simulation needs to be logically structured, to better help stakeholders comprehend the requirements.

Stakeholders: Seismologists, high performance computing experts, software engineers and risk management organizations.

Problem: How to specify various required elements for a forward simulation process. A forward simulation process has many requirements and some might be forgotten at an initial project stage or lead to misunderstanding between various stakeholders. Stakeholders need to be able to easily communicate and exchange their expertise, to achieve a common agreement on the requirements.

Forces:

- Various types of data, which describe seismic waves and earth properties, should be given as input data for the forward simulation. Data types are application-specific, as well as how data is represented.
- The complexity of numerical calculation vs. easy-to-control: Numerical calculations in seismology are often complex and involve sophisticated numerical operations. Numerical methods that are used in forward simulation influence calculation and its results greatly. A calculation procedure often needs to be

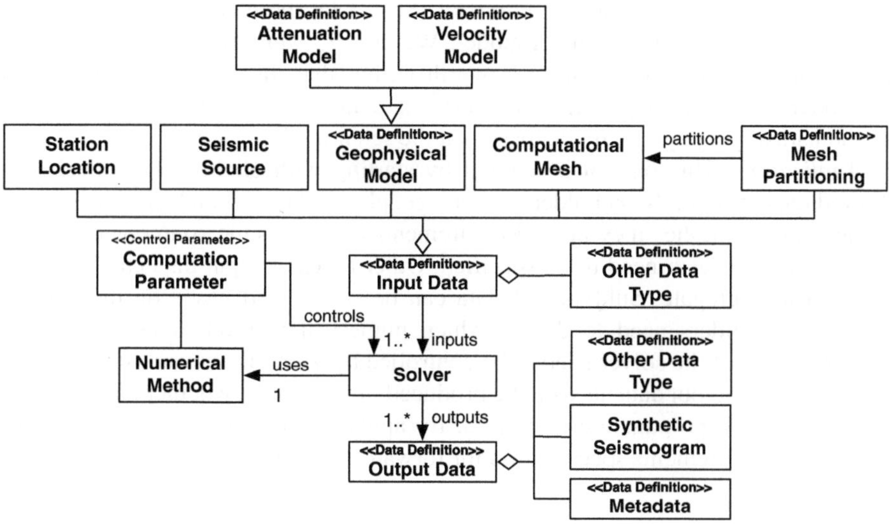

Fig. 7 UML class diagram of the forward simulation pattern

adjusted by users to achieve specific numerical requirements based on the seismological problem to solve and the required numerical precision. Users also expect to easily control the calculation to output required types of data.

- Representation of output data must be well defined, for example output according to a data format standard.

Solution: An aggregate of input data should be given to a forward simulation solver. The pattern provides generic types of data that are commonly required in forward simulation input. Special types of data for intended applications can be specialized or extended based on the pattern. Computation parameters should be specified to control the numerical calculation. For example, the time step of the simulation and the order of accuracy of the numerical method should be set. The pattern also indicates that stakeholders should agree on the requirements of output data.

Figure 7 shows elements and their relations of the forward simulation pattern in a Unified Modeling Language (UML) class diagram. The InputData for a forward simulation Solver usually consists of StationLocation, SeismicSource, GeophysicalModel, and ComputationalMesh. To perform parallel computing, the computational mesh needs to be partitioned. AttenuationModel and VelocityModel are two common types of geophysical models used in a forward simulation process. A Solver does main calculation for the forward simulation based on a specific numerical method and outputs calculation results. ComputationParameter controls the solver. The most common output data is SyntheticSeismogram. Meanwhile, the Metadata, for instance about the synthetic seismogram format or about the location information, is also generated.

Referring to the extended meta-model (Sect. 3.1), the elements of the pattern are of defined types in the meta-model. We use ≪ *stereotype* ≫ to indicate a type of a specialized element. It is worth noting that the geophysical model in the pattern is of the type *data definition*. It denotes the data that describe the mathematical/geophysical model based on the scientific knowledge.

Application: This pattern is applied by starting with eliciting and specifying desired data to input. Stakeholders select a geophysical model and specify types of computational meshes they need. Requirements about seismic source and seismic station locations also need to be specified. Other types of input data can be added to the data aggregate. Subtypes of data can be specialized based on the generic types. Stakeholders need to decide which numerical method to use, and what functionalities of the solver need to be controlled by users. In the end, stakeholders agree on what output data needs to be produced, for example, whether an animation of seismic wave propagation is desired. By applying this pattern, a requirements model or a requirements specification for forward simulation is created.

3.2.2 Data Access Pattern

Data access, data mining, data transfer and data storage are important issues in seismology software projects. Seismology related data are stored in different media and in multiple locations. Sets of compatible data from worldwide stations and networks over time are collected in data facilities (e.g. IRIS and ORFEUS). These data are crucial to providing reliable results in seismology applications.

Name: Data Access Pattern

Activity: The pattern can be used during requirements elicitation and specification activities, when data needs to be accessed, especially to be accessed externally.

Stakeholders: Seismologists, high performance computing experts, software engineers, database experts and risk management organizations.

Problem: How to access different types of data in a seismology software application. Sometimes, developers face implementation difficulties that some data cannot be accessed easily. Therefore, they might need to redesign the software architecture to incorporate such issues. This problem should be mitigated by considering constraints of access data during requirements elicitation.

Forces:

- Stakeholders might want to integrate data access into a system. However, different types of data have different interfaces and regulations for access.
- Users should be able to interactively explore the observed and synthetic seismograms.
- Data should be able to be transferred between HPC facilities to enable simulation and post-process the calculated results.
- Data formats, storage and exchange standards are not yet fully established or not well applied in the seismology community.

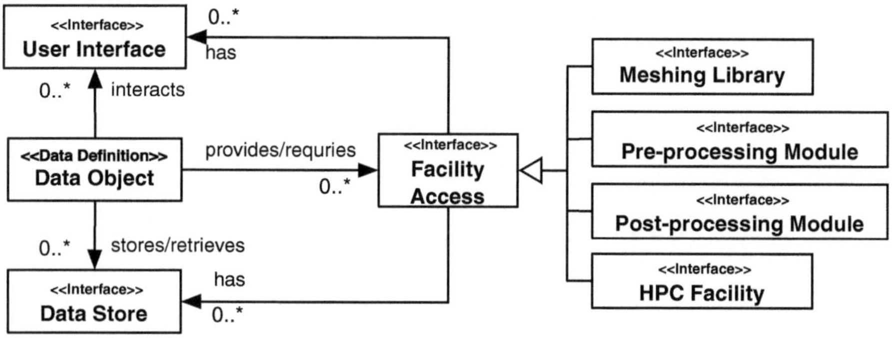

Fig. 8 UML class diagram of the data access pattern

Solution: For each type of data object to access, interfaces to the data access should be specified explicitly and individually. Figure 8 shows the UML class diagram for the data access pattern. DataObject can be required from interfaces of Facilities. On the other hand, DataObject can also provide interfaces for access. For instance, an external visualization tool can visualize output data of a system by means of the provided data access. Facilities such as MeshTool, Pre-processingTool, Post-processingTool and HPCFacility often provide or require access to data objects. Sometimes, it is necessary to interact with data through a UserInterface. For example, a user needs to select a certain region of earth via a user interface, in order to generate a computational mesh. Data can be stored or retrieved from different types of Data Store.

Application: Users apply the pattern to specify a type of data object that requires external access from facilities outside or provides access to the outside. For example, computational meshes of required types might need to be generated by certain external meshing tools. Stakeholders also need to decide whether a user interface is required or data needs to be stored and in what form. Constraints of access need to be specified clearly. For instance, licensing issues of external tools, access protocols to HPC systems need to be defined.

3.3 Implementation

We have implemented a prototypical tool of requirements modeling tailored for seismological software development. The tool is developed as a set of Eclipse plug-ins. It is based on the extended meta-model. Users can create a requirements model instance for a seismology project. This instance contains model elements of various types that are predefined in the meta-model. The contained model elements can be linked according to the predefined relationships between types. The tool also supports utilizing the identified patterns.

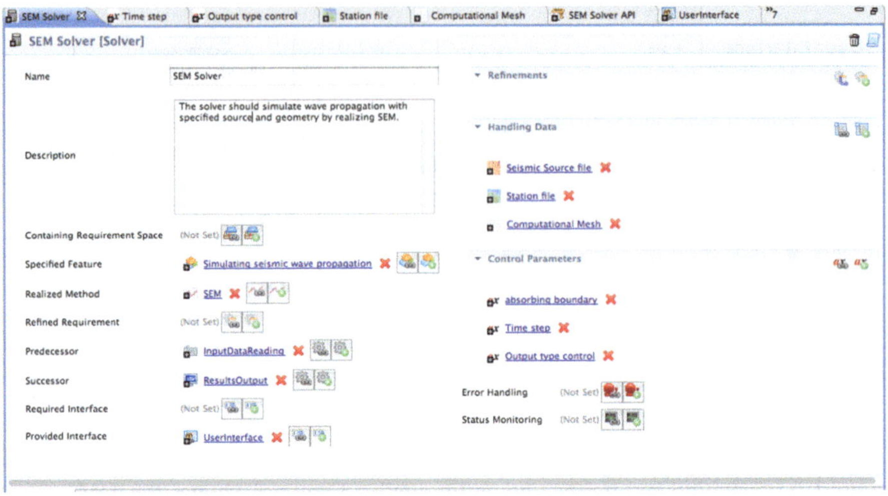

Fig. 9 Modeling requirements for the SEM solver in the editor

Figure 9 illustrates modeling requirements for SEM solver in the editor of our tool. A requirement of the solver type, "the solver should simulate wave propagation with specified source and geometry by realizing SEM", is created. Developers specify its realized numerical method, predecessor and successor. A user interface should be provided to allow users easily use the SEM solver. Handling data as well as control parameters for the SEM solver are specified. Developers jump into the editing page of the hyperlinked elements and detail them.

4 Example Scenario

This section presents an example scenario to illustrate how the extended meta-model, as well as the identified patterns are applied in a system of dynamic rupture simulation. Dynamic rupture is a source type of particular interest for earthquake engineering and seismic hazard assessment [19]. Based on the meta-model, seismologists first quickly brainstorm and organize their scientific knowledge about dynamic rupture simulation. The scientific problem is combining earthquake rupture and wave propagation in complex fault geometries. Tetrahedra are used to better represent the geometrical constrains of a fault[2] and a friction law is used to model initial rupture. The medium is assumed to be isotropic. A high-order discontinuous Galerkin (DG) method combined with an arbitrarily high-order derivatives (ADER)

[2] A surface along an earthquake rupture is propagating

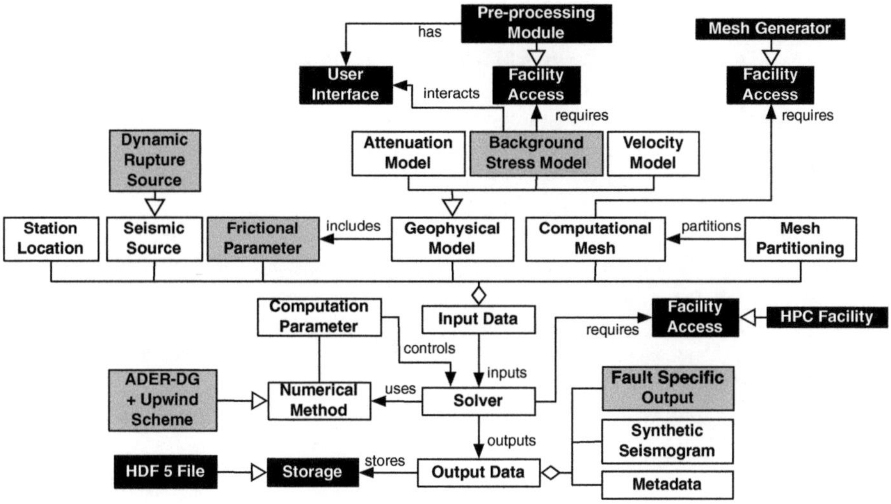

Fig. 10 Simplified UML class diagram of dynamic rupture

time integration method is employed to solve the three-dimensional elastodynamic equations.

To computationally solve this problem, seismologists want to develop a feature of simulating dynamic rupture using ADER-DG scheme. They specify the hardware, on which the simulation runs. They further detail requirements for this feature, where the two identified patterns are applied to help them specify required elements productively.

Figure 10 is a simplified class diagram of the requirements for the dynamic rupture simulation application. The gray elements in the diagram indicate that they are specified or newly added requirements based on the forward simulation pattern. The seismic source is now specialized to be a *dynamic rupture source* type. In addition to attenuation models and velocity models, the application should also support *background stress models*. Furthermore, the geophysical models should include *frictional parameters* to represent the status of the earthquake fault. It should use tetrahedral elements as *computational meshes* and the given *station locations* should be converted to UTM coordinates.[3] The simulation of dynamic rupture prefers a particular solver namely the *ADER-DG method in combination with Upwind Schemes*. As for the output, it should generate *fault specific output* besides synthetic seismograms. The application should also output metadata that consists of information about the solver, the input mesh, the seismic source and the geophysical model.

[3]Universal Transverse Mercator (UTM) is a geographic coordinate system, which uses a 2-dimensional Cartesian coordinate system to give locations on the surface of the Earth.

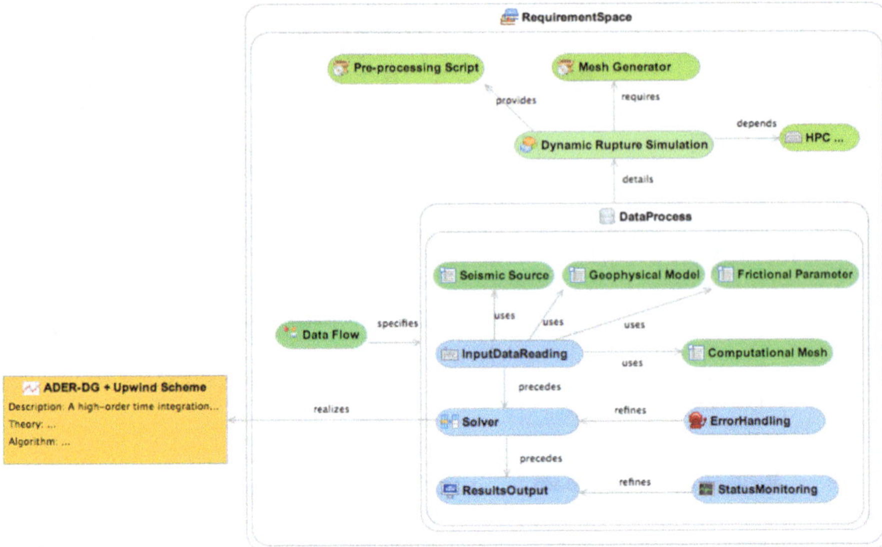

Fig. 11 Graphically creating a requirements model of dynamic rupture in the prototypical tool

The black elements are created employing the data access pattern. The background stress model needs to be *pre-processed* via a *user interface*. The solver should be able to access an *HPC facility* to transfer calculation data, where valid user accounts on the HPC with suitable privilege should be organized. An external *mesh generator* should be used to generate the required two- and three-dimensional meshes. The output data should be stored in *HDF 5 files*.

Using the tool prototype we introduced in Sect. 3.3, we can also create the requirements model graphically (see Fig. 11). Each model element (a node) can be elaborated into details. It takes about 10 min for a developer to create such an initial model. The model can be further refined based on group review and discussion. Through the mentioned requirements engineering practice, seismologists have clearer ideas about what they want for the feature. They are able to create and specify requirements in an efficient way without much pre-knowledge/training in requirements engineering. In particular, the requirements elicitation process is more structured based on the meta-model and patterns. Discussion about issues such as how to access certain data is carried out to support decision-making on the software design.

5 Related Work

There are many requirements modeling techniques that help specify and analyze requirements, such as use case and goal-oriented techniques. These techniques need to be properly selected to use in the context of the project for different needs. Use

Case is commonly used to describe a flow of events and the interaction between the system and the user [5]. However, it is not suitable to capture self-contained requirements that are not user-interaction based (e.g. a required mathematical calculation algorithm) and non-functional requirements (e.g. the desired computing performance and efficiency), which are the majority of the requirements in a seismological software project. Goal-oriented approaches support heuristic, qualitative or formal reasoning schemes during requirements engineering [20]. Goals provide the rationale, which provides a high-level view of the system to support decision-making and help derive low-level details. However, using goal-oriented approaches requires extensive learning effort, which is less likely to achieve in a seismological research project, due to time and expertise limitation.

The approach we proposed utilizes domain knowledge to provide a more easy-acceptable and applicable environment for requirements modeling. In terms of visual modeling, the requirements can be represented in an UML class diagram. Therefore, it can be easily implemented using existing technologies. It can also be transformed to the object model during software design.

6 Conclusion and Future Work

This paper discussed the importance of requirements in seismology software development and what requirements engineering practices are needed. To address the problem of missing or incomplete requirements, we introduced a model-based approach to easily capture and manage requirements. The approach is based on a customizable meta-model and a semi-automated domain literature analysis process, to reuse domain and software development knowledge in the computational seismology domain. We identified two requirements patterns in order to define requirements in a more efficient and effective manner, by providing structured knowledge from successful software projects. We have implemented a prototypical tool to support requirements engineering for computational seismology, based on the customized meta-model and identified patterns.

We are already working on evaluating the approach in various seismological projects, to further measure how efficient the approach is in requirements engineering practices and investigate the quality of created requirements. On the other hand, we are still improving automation of text analysis by employing more advanced computational linguistic algorithms and creating more training data. In the future, we will identify other requirements patterns for different seismological software applications such as probabilistic seismic hazard assessment tools commonly used to support earthquake risk management. Another interesting extension of this work is to provide a catalog that consists of identified patterns of the domain. We will also adapt the approach in a different scientific domain such as computational physics and examine its applicability.

Acknowledgements We would like to thank all model reviewers, for their valuable feedback to help us refine the proposed model. This work is supported by Munich Centre of Advanced Computing (MAC).

References

1. Ackroyd, K.S., Kinder, S.H., Mant, G.R., Miller, M.C., Ramsdale, C.A., Stephenson, P.C.: Scientific software development at a research facility. Softw IEEE **25**(4), 44–51 (2008)
2. Chandrasekaran, B., Josephson, J., Benjamins, V.: What are ontologies, and why do we need them? Intell. Syst. Appl. IEEE **14**(1), 20–26 (1999). doi:10.1109/5254.747902
3. Cheng, B.H.C., Atlee, J.M.: Research directions in requirements engineering. In: 2007 Future of Software Engineering, FOSE '07, pp. 285–303. IEEE Computer Society, Washington, DC (2007). doi:http://dx.doi.org.eaccess.ub.tum.de/10.1109/FOSE.2007.17. http://dx.doi.org.eaccess.ub.tum.de/10.1109/FOSE.2007.17
4. CIG: Specfem3d. http://www.geodynamics.org/cig/software/specfem3d (2013)
5. Cockburn, A.: Writing Effective Use Cases, vol. 1. Addison-Wesley, Boston (2001)
6. Gómez-Pérez, A., Benjamins, R.: Overview of knowledge sharing and reuse components: ontologies and problem-solving methods. In: IJCAI and the Scandinavian AI Societies. CEUR Workshop Proceedings (1999)
7. Gruber, T.R.: A translation approach to portable ontology specifications. Knowl. Acquis. **5**(2), 199–220 (1993). doi:10.1006/knac.1993.1008. http://dx.doi.org.eaccess.ub.tum.de/10.1006/knac.1993.1008
8. Hannay, J.E., MacLeod, C., Singer, J., Langtangen, H.P., Pfahl, D., Wilson, G.: How do scientists develop and use scientific software? In: 2009 ICSE Workshop on Software Engineering for Computational Science and Engineering, pp. 1–8 (2009). doi:10.1109/SECSE.2009.5069155. http://ieeexplore.ieee.org/lpdocs/epic03/wrapper.htm?arnumber=5069155
9. Helming, J., Koegel, M., Schneider, F., Haeger, M., Kaminski, C., Bruegge, B., Berenbach, B.: Towards a unified requirements modeling language. In: 2010 Fifth International Workshop on Requirements Engineering Visualization (REV), pp. 53–57 (2010). doi:10.1109/REV.2010.5625659
10. Jackson, M.: Some notes on models and modelling. In: Borgida, A.T., Chaudhri, V.K., Giorgini, P., Yu, E.S. (eds.) Conceptual Modeling: Foundations and Applications, pp. 68–81. Springer, Berlin/Heidelberg (2009)
11. Kaeser, M.: Seissol: high resolution simulation of seismic wave propagation. http://www.geophysik.uni-muenchen.de/~kaeser/SeisSol/. (2013)
12. Kaiya, H., Saeki, M.: Using domain ontology as domain knowledge for requirements elicitation. In: 14th IEEE International Conference on Requirements Engineering, pp. 189–198 (2006). doi:10.1109/RE.2006.72
13. Konrad, S., Cheng, B.H.: Requirements patterns for embedded systems. In: IEEE International Conference on Requirements Engineering, vol. 0, p. 127 (2002). doi:http://doi.ieeecomputersociety.org/10.1109/ICRE.2002.1211541
14. Li, Y.: Reengineering a scientific software and lessons learned. In: Proceedings of the 4th International Workshop on Software Engineering for Computational Science and Engineering, SECSE '11, pp. 41–45. ACM, New York (2011). doi:10.1145/1985782.1985789. http://doi.acm.org.eaccess.ub.tum.de/10.1145/1985782.1985789
15. Li, Y., Harutunian, M., Narayan, N., Bruegge, B., Buse, G.: Requirements engineering for scientific computing: a model-based approach. In: 2011 IEEE Seventh International Conference on e-Science Workshops (eScienceW), pp. 128–134 (2011). doi:10.1109/eScienceW.2011.30
16. Li, Y., Narayan, N., Helming, J., Koegel, M.: A domain specific requirements model for scientific computing: nier track. In: Proceeding of the 33rd International Conference on Software Engineering, ICSE '11, Honolulu, pp. 848–851. ACM, New York (2011)

17. Li, Y., Pelties, C., Kaser, M., Nararan, N.: Requirements patterns for seismology software applications. In: 2012 IEEE Second International Workshop on Requirements Patterns (RePa), pp. 12–16 (2012). doi:10.1109/RePa.2012.6359967
18. Maynard, D., Li, Y., Peters, W.: Nlp techniques for term extraction and ontology population. In: Proceedings of the 2008 Conference on Ontology Learning and Population: Bridging the Gap between Text and Knowledge, pp. 107–127. IOS Press, Amsterdam (2008). http://dl.acm.org.eaccess.ub.tum.de/citation.cfm?id=1563823.1563834
19. Pelties, C., la Puente, J.D., Ampuero, J.P., Brietzke, G.B., Käser, M.: Three-dimensional dynamic rupture simulation with a high-order discontinuous galerkin method on unstructured tetrahedral meshes. J. Geophys. Res. Solid Earth **117** (2012). doi:10.1029/2011JB008857
20. Van Lamsweerde, A.: Goal-oriented requirements engineering: a guided tour. In: Proceedings of the Fifth IEEE International Symposium on Requirements Engineering, 2001, Toronto, pp. 249–262. IEEE (2001)
21. VERCE: Virtual earthquake and seismology research community e-science environment in europe. www.verce.eu (2013)
22. Wen, Y., Zhao, H., Liu, L.: Analysing security requirements patterns based on problems decomposition and composition. In: 2011 First International Workshop on Requirements Patterns (RePa), pp. 11–20 (2011). doi:10.1109/RePa.2011.6046726

A High-Performance Interactive Computing Framework for Engineering Applications

Jovana Knežević, Ralf-Peter Mundani, and Ernst Rank

Abstract To harness the potential of advanced computing technologies, efficient (real time) analysis of large amounts of data is as essential as are front-line simulations. In order to optimise this process, experts need to be supported by appropriate tools that allow to interactively guide both the computation and data exploration of the underlying simulation code. The main challenge is to seamlessly feed the user requirements back into the simulation. State-of-the-art attempts to achieve this, have resulted in the insertion of so-called check- and break-points at fixed places in the code. Depending on the size of the problem, this can still compromise the benefits of such an attempt, thus, preventing the experience of real interactive computing. To leverage the concept for a broader scope of applications, it is essential that a user receives an immediate response from the simulation to his or her changes. Our generic integration framework, targeted to the needs of the computational engineering domain, supports distributed computations as well as on-the-fly visualisation in order to reduce latency and enable a high degree of interactivity with only minor code modifications. Namely, the regular course of the simulation coupled to our framework is interrupted in small, cyclic intervals followed by a check for updates. When new data is received, the simulation restarts automatically with the updated settings (boundary conditions, simulation parameters, etc.). To obtain rapid, albeit approximate feedback from the simulation in case of perpetual user interaction, a multi-hierarchical approach is advantageous. Within several different engineering test cases, we will demonstrate the flexibility and the effectiveness of our approach.

Keywords Interactive computing • CSE applications • High-performance computing

J. Knežević (✉) · R.-P. Mundani · E. Rank
Technische Universität München, Arcisstr. 21, 80333 München, Germany
e-mail: knezevic@bv.tum.de; mundani@tum.de; ernst.rank@tum.de

M. Bader et al. (eds.), *Advanced Computing*, Lecture Notes in Computational Science and Engineering 93, DOI 10.1007/978-3-642-38762-3_9,

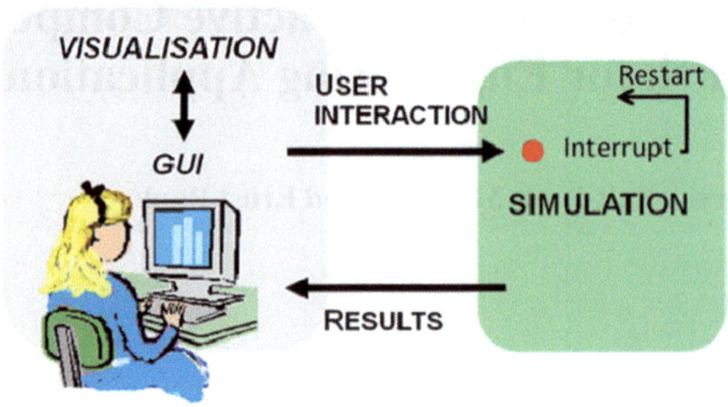

Fig. 1 A user guides an often time and memory consuming simulation in order to build a solution to his/her problem via a graphical user interface

1 Introduction

Simulation of very complex physical phenomena becomes a realistic endeavour with the latest advances in hardware technologies, sophisticated (numerical) algorithms, and efficient parallelisation strategies. It consists of modelling a domain of a physical problem, applying appropriate boundary conditions, and doing numerical approximation for the governing equations, often with a linear or non-linear system as outcome. When the system is solved, the result is validated and visualised for more intuitive interpretations.

All the aforementioned cycles – pre-processing, computation, and post-processing – can be very time consuming, depending on the discretisation parameters, e.g., and moreover, are traditionally carried out as a sequence of steps. The ever-increasing range of specialists in developing engineering fields has necessitated an interactive approach with the computational model. This requires real-time feedback from the simulation during program runtime, while experimenting with different simulation setups. For example, the geometry of the simulated scene can be modified interactively altogether with boundary conditions or a distinct feature of the application, thus, the user can gain "insight concerning parameters, algorithmic behaviour, and optimisation potentials" [23].

Interactive computing frameworks, libraries, and Problem Solving Environments (PSEs) are used by specialists to interact with complex models, while not requiring deep knowledge in algorithms, numerics, or visualisation techniques. These are user-friendly facilities for guiding the numerically approximated problem solution. The commonly agreed features are: a sophisticated user interface for the visualisation of results on demand and a separated steerable, often time- and memory-consuming simulation running on a high-performance computer (see Fig. 1).

The concept has been present in the scientific and engineering community already for more than two decades. Meanwhile, numerous powerful tools serving this purpose have been developed. A brief overview of some state-of-the-art tools – steering environments and systems such as CSE [33], Magellan [35], SCIRun [26], Uintah [4], G-HLAM [29] and EPSN [25] , libraries such as

CUMULVS [9], or RealityGrid [3, 27], or frameworks such as Steereo [12] – is provided in the next section. Those tools differ in the way they provide interactive access to the underlying simulation codes, using check- and breakpoints, satellites connected to a data manager, or data flow concepts, e.g., hence they cannot always fully exploit interactive computing and are usually of limited scope concerning different application domains.

2 Computational Steering: State of the Art

CSE [33] is a computational steering environment consisting of a very simple, flexible, minimalistic kernel and modular components, so-called satellites, where all the higher level functionality is pushed. It is based on the idea of a central process, i.e. a data manager to which all the satellites can be connected. Satellites can create and read/write variables, and they can subscribe to events such as notification of mutations of a particular variable [22]. The data manager informs all the satellites of changes made in the data and an interactive graphics editing tool allows users to bind data variables to user interface elements.

CUMULVS [9] is a library that provides steering functionality so that a programmer can extract data from a running (possibly parallel) simulation and send those data to the visualisation package. It encloses the connection and data protocols needed to attach multiple visualisation and steering components to a running application during execution. The user has to declare in the application which parameters are allowed to be modified or steered, or the rules for the decomposition of the parallel data, etc. Using check-pointing, the simulation can be restarted according to the new settings.

In the steering system called Magellan [35], steering objects are exported from an application. A collection of instrumentation points, such as so-called actuators, know how to change an object without disrupting application execution. Pending update requests are stored in a shared buffer until an application thread polls for them [35].

EPSN [25] API is a distributed computational steering environment, where an XML description of simulation scripts is introduced to handle data and concurrency at instrumentation points. There is a simple connection between the steering servers, i.e. simulation back ends, and clients, i.e. user interfaces. When receiving requests, the server determines their date, thus, the request is executed as soon as it fulfills a condition. Reacting on a request means releasing the defined blocking points.

Steereo [12] is a light-weight steering framework, where the client can send requests and the simulation will respond to them. However, the requests are not processed immediately, but rather stored in a queue and executed at predefined points in the simulation code. Hence, users have to define when and how often this queue should be processed.

The RealityGrid [3, 27] project has provided a highly flexible and robust computing infrastructure for supporting the modelling of complex systems [10].

An application is structured into a client, a simulation, and a visualisation unit communicating via calls to the steering library functions. Also this infrastructure involves the insertion of check- and break-points at fixed places in the code where changed parameters are obtained and the simulation is to be restarted.

In the SCIRun [26] problem solving environment (PSE) for modelling, simulation, and visualisation of scientific problems, a user may smoothly construct a network of required modules via a visual programming interface. Computer simulations can then be executed, controlled, and tuned interactively, triggering the re-execution only of the necessary modules, due to the underlying dataflow model. It allows for extension to provide real-time feedback even for large scale, long-running, data-intensive problems. This PSE has typically been adopted to support pure thread-based parallel simulations so far. Uintah [4] is a component-based visual PSE that builds upon the best features of the SCIRun PSE, specifically addressing the massively parallel computations on petascale computing platforms.

In the G-HLAM [29] PSE, the focus is more on fault tolerance, i.e. monitoring and migration of the distributed federates. The group of main G-HLAM services consists of one which coordinates management of the simulation, one which decides when performance of a federate is not satisfactory and migration is required, the other which stores information about the location of local services. It uses distributed federations on the Grid for the communication among simulation and visualisation components.

All of those powerful tools have, however, either limited scope of application, or are involving major simulation code changes in order to be effective. This was the motivation for us to design a new framework that incorporates the strong aspects of the aforementioned tools, nevertheless overcomes their weak aspects in order to provide a generic concept for a plenitude of different applications with minimal codes changes and a maximum of interactivity.

Within the Chair for Computation in Engineering at Technische Universität München, a series of successful Computational Steering research projects took place in the previous decade. It has also involved collaboration with industry partners. Performance analysis has been done for several interactive applications, in regard to responsiveness to steering, and the factors limiting performance have been identified. The focus at this time was on interactive computational fluid dynamics (CFD), based on the Lattice-Boltzmann method, including a Heating Ventilation Air-Conditioning (HVAC) system simulator [2], online-CFD simulation of turbulent indoor flows in CAD-generated virtual rooms [37], interactive thermal comfort assessment [34], and also on structure mechanics – computational methods in orthopaedics. Over time, valuable observations and experience have resulted in significant reduction of the work required to extend an existing application code for steering.

Again, the developed concepts have been primarily adapted to this limited number of application scenarios, thus, they allow for further investigations so as to become more efficient, generic, and easy to implement. This is where our framework comes into play.

3 The Idea of the Framework

For widening the scope of the steerable applications, an immediate response of any simulation back end to the changes made by the user is required. Hence, the regular course of the simulation has to be interrupted as soon as a user interacts. Within our framework, we achieve this using the software equivalent of hardware interrupts, i.e. signals. The check for updates is consequently done in small, user-defined, cyclic intervals, i.e., within a function handling the Unix ALARM signal.

If the check does not indicate any update from the user side, the simulation gets the control back and continues from the state saved at the previous interrupt-point. Otherwise, the new data is received, matched to the simulation data (which is the responsibility of the user himself), and simulation state variables (for instance loop delimiters) are manipulated in order to make the computation stop and then automatically start anew according to the user modifications. Taking the pseudo code of an iteratively executed function (within several nested loops) as an example, the redundant computation is skipped as soon as the end of the current, most-inner loop iteration is reached. This is, namely, the earliest opportunity to compare the values of the simulation state variables, and, if the result of the comparison indicates so, exit all the loops (i.e. starting with most-inner one and finishing with the most-outer one) [13–15]. This would exactly mean starting computation over again, as illustrated in the pseudocode:

```
% X_end, Y_end declared global

begin function Signal_Handler()
  % manipulate X_end, Y_end so that the redundant
  % computation is skipped and started anew
  X_end = Y_end = -1
end

% set time interval for periodically occurrence of
% ALARM signal to stop execution and call handler
Set_Alarm()

% user function to be interrupted
begin function Compute()
  for t = T_start to T_end do
    for i = X_start to X_end do
      for j = Y_start to Y_end do
        Process(data[i][j])
        % potential update is recognised next
      od
    od
  od
end
```

As elaborated in [13], to guarantee the correct execution of a program, one should use certain type qualifiers (provided by ANSI C, e.g.) for the variables which are subject to sudden change or objects to interrupts. One should ensure that certain types of objects which are being modified both in the signal handler and the main computation are updated in an atomic way. Furthermore, if the value in the signal handler has changed, the outdated value in the register should not be used again. Instead, the new value should be loaded from memory. This may occur due to the custom compiler optimisations. In addition to this, sufficient steps have to be taken to prevent potentially introduced severe memory leaks before the new computation is started. This is due to the interrupts and their possible occurrence before the memory allocated at a certain point has been released.

Finally, with either one or several number of iterations being finished without an interrupt, new results are handed on to the user process for visualisation. One more time it is the user's responsibility to prescribe to the front end process how to interpret the received data so that it can be coherently visualised [13–16, 18].

In Fortran, similar to C/C++, support for signal handling can be enabled at user level with minimal efforts involved. Some vendor supplied Fortran implementations, including for example Digital, IBM, Sun, and Intel, have the extension that allows the user to do signal handling as in C [1]. Here, a C wrapper function for overriding the default signal behaviour has to be implemented. However, the behaviour of the Fortran extension of the aforementioned function is implementation dependent, and if the application is compiled using an Intel Fortran compiler, when the program is interrupted, it will terminate unless one "clears" the previously defined action first.

Due to the accuracy requirements and the increasing amount of data which has to be handled in numerical simulations of complex physical phenomena nowadays, there is an urge to fully exploit the general availability and increasing CPU power of high-performance computers. For this, in addition to efficient algorithms and data structures, sophisticated parallel programming methods are a constraint. The design of our framework, therefore, takes into consideration and supports different parallel paradigms, which results in an extra effort to ensure correct program execution and avoid synchronisation problems when using threads, as explained in the following subsection.

3.1 Multithreading Parallelisation Scenario

We consider the scenario when pure multithreading (with, e.g., OpenMP/POSIX threads) is employed in the computations on the simulation side. Since a random thread is interrupted via signal at the expiration of the user-specified interval, that thread probes, via the functionality of the Message Passing Interface (MPI), if any information regarding the user activity is available. If the aforesaid checking for a user's message indicates that an update has been sent, the receiving thread instantly obtains all the information and applies necessary manipulations in order to re-start the computation with the changed setting. Hence, all other threads also

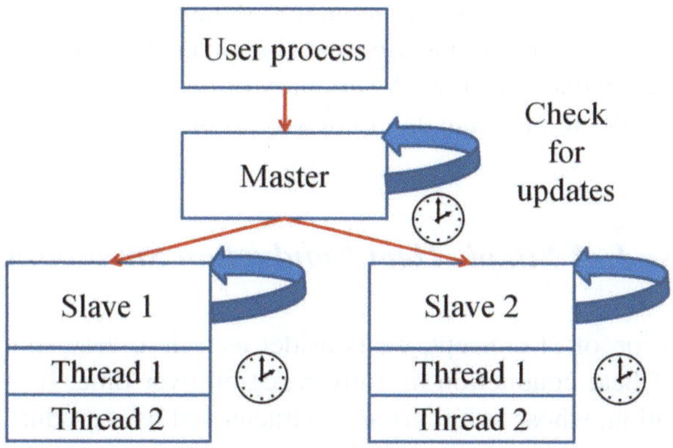

Fig. 2 In the case of a hybrid parallel scenario, each process is doing its own checks for updates; one random thread per process is interrupted in small, fixed time intervals

become instantly aware that their computations should be started over again and must now proceed in a way in which clean termination of the parallel region is guaranteed.

3.2 "Hybrid" Parallelisation Scenario

In a "hybrid" parallel scenario (i.e. MPI and OpenMP – see Fig. 2), a random thread in each active process is being interrupted, hence, fetches an opportunity to check for the updates. The rest of the procedure is similar as described for the pure multithreading, except that now all the processes have to be explicitly notified about the changes performed by a user. This may involve additional communication overheads. Moreover, if one master process, which is the direct interface of the user's process to the computing-nodes, i.e. slaves, is supposed to inform all of them about the user updates, it may become a bottleneck. Therefore, a hierarchical non-blocking broadcast algorithm for transferring the signal to all computing nodes has been proposed in [14, 15].

4 Applications

In the following, a few application scenarios are presented, where the implemented framework has been successfully integrated. First, a simple 2D simulation of a temperature conduction, used only for testing purposes, where heat sources, boundaries of the domain, etc. can be interactively modified. Then, a neutron transport simulation developed at the Nuclear Engineering Program, University of Utah, which has been the first Fortran test case for the framework. The next one is the sophisticated Problem Solving Environment SCIRun developed at the Scientific Computing and Imaging (SCI) Institute, University of Utah. The final one is a tool for pre-operative planning of hip-joint surgeries, done as a collaborative project of the Chair for Computation in Engineering and Computer Graphics and Visualization at Technische Universität München. A summary of necessary modifications of the original codes in order to integrate the framework is discussed in Sect. 5.2.

4.1 Test Case 1: A Simple Heat Conduction Simulation

Simulation: For proof of concept, we consider as a first, very simple example a 2D simulation of heat conduction in a given region over time. It is described by the Laplace equation, whose solutions are characterised by a gradual smoothing of the starting temperature distribution by the heat flow from warmer to colder areas of a domain. Thus, different states and initial conditions will tend toward a stable equilibrium. After numerical treatment of the PDE via a Finite Difference scheme, we come up with a five-point stencil. The Gauss-Seidel iteration method is used to solve the resulting linear system of equations.

GUI/Visualisation: For interacting with the running simulation, a graphical user interface is provided using the wxWidgets library [11]. The variations of the height along the z-axes, pointing upward, are representing the variations of the temperature in the corresponding 2D domain. Both the simulation and the visualisation are implemented in C++ and are separate MPI processes.

User interaction: When it comes to the interplay with the program during the simulation, there are a few possibilities available – one can interactively add, delete, or move heat sources, add, delete, or move boundary points of the domain, or change the termination condition (maximal number of iterations or error tolerance) of the solver. As soon as a user interacts, the simulation becomes immediately aware of it and consequently the computation is restarted. An instant estimation of the equilibrium state for points of the domain far away from the heat sources is unfortunately not always feasible on the finest grid used (300×300). This may be the case due to the short intervals between two restarts in case of too frequent user interaction, as shown in Fig. 3. Here we profit from a hierarchical approach.

The **hierarchical approach** is based on switching between several grids of different resolutions depending on the frequency of the user interaction. At the beginning, the finest desired grid is used for the computation. When the simulation process is interrupted by an update, it restarts the computation with the new settings, but on a coarser grid for faster feedback, i.e. to provide new results as soon as possible. As long as the user is frequently interacting, all computations are carried out on the coarser grids only. If the user halts, i.e. stops interacting, the computation switches back to the finest grid in order to provide more accurate values. In this particular test case, three different grids were used – an initial 300×300 grid, the four times smaller, intermediate one (150×150, in case of lower pace of interactions, i.e. adding/deleting heat sources or boundary points, e.g.) and, finally, the coarsest

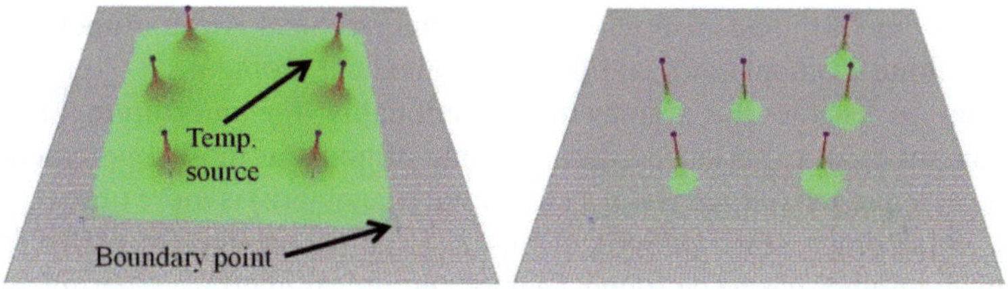

Fig. 3 *Left*: an initial scenario; *right*: moving heat sources/boundaries leads to the restart of the computation, but in-between two too frequent restarts a user is unable to estimate the equilibrium temperature in the region farther away from the heat sources (here, further iterations of the solver would be necessary)

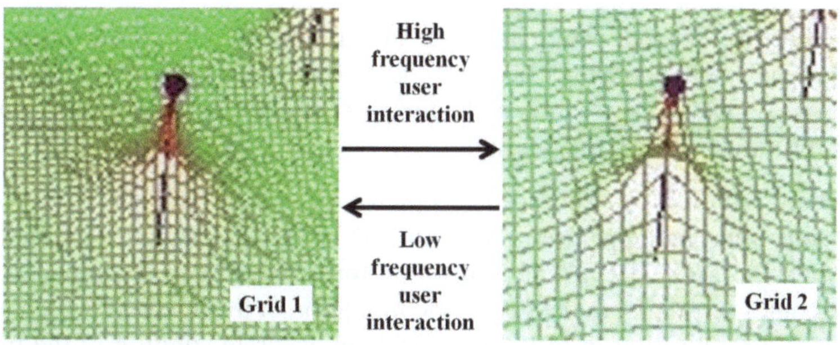

Fig. 4 Switching to a coarser grid in case of moving heat sources or boundaries, switching back to the finer one once the user stops interacting

one (75×75) for very high frequency moving of boundary points or heat sources over the domain (Fig. 4). The coarser grids are not meant for obtaining quantitative solutions, just for a fast qualitative idea how the solution might look like. If a user interactively found an interesting setup, he just has to stop and an accurate solution for this setup will be computed. Nevertheless, measurements concerning the different grids showed that the variation of the solution on the finest grid compared to the intermediate one is around 4.5 %, and compared to the coarsest one around 14.6 %. The described approach, on the other hand, leads to an improvement in convergence by a factor of 2.

Additionally, we employ a multi-level algorithm – the results of the computation on the coarsest grid are not discarded when switching to the finer one. Our concept, namely, already involves a hierarchy of discretisations, as is the case in multigrid algorithms, thus, we can profit from the analogous idea. Our scheme is somewhat simpler – it only starts with the solution on the coarsest grid and uses the result we gain as an initial guess of the result on a finer one. A set of examples has been tested (with grids 300×300, 150×150, and 75×75) where the number of necessary operations on the intermediate and fine grid could be halved. What seems to be a somehow obvious approach, at least for this simple test scenario, can be efficiently exploited in our forth test case where we use hierarchical Ansatz functions for an interactive finite-element computation of a biomedical problem.

In order to enable the framework functionality for interrupting the above simulation, takes an experienced user a couple of hours at most. Implementing the hierarchical approach (which is not a part of the framework) is more time consuming (a few working days), since an optimal automatic detection when to switch from one hierarchy to another has to be found—which requires numerous experiments.

4.2 Test Case 2: A Neutron Transport Simulation

We present as second test case the integration of the framework into a computationally efficient, high accuracy, geometry independent neutron transport simulation. It makes researchers' and educators' interaction with virtual models of nuclear reactors or their parts possible.

Simulation: AGENT (Arbitrary GEometry Neutron Transport) solves the Boltzmann transport equation, both in 2D and 3D, using the Method of Characteristics (MOC) [21]. The motivation for steering such a simulation during runtime comes mostly from the geometric limitation of this method, which requires fine spatial discretisation in order to provide an accurate solution to the problem. On the other hand, a good initial solution guess would help tremendously to speed-up the convergence, and this property is used to profit from our framework. The 3D discretisation basis for the Boltzmann equation consists of the discrete number of plains, for each of which both a regular geometry mesh and a number of parallel rays in a discrete number of directions are generated. The approximation results in a system of equations to be iteratively solved for discrete fluxes.

GUI/Visualisation: The result in terms of the scalar fluxes is simultaneously calculated and periodically visualised. The simulation server maintains a list of available simulation states, and clients connect using the ImageVis3D volume rendering tool [8] to visualise the results in real time. Users can interfere with the running simulation via a simple console interface, providing the new values of the desired parameters.

User interaction: Instant response of the simulation to the changes made by the user is again achieved via signals. Using the technique described as our general concept, the most outer iteration instantly starts anew, as soon as its overall state is reset within the main computational steering loop, according to the updated settings and necessary re-initialisation of the data. By manipulating only two simulation parameters in the signal handler, it is achieved that the iteration restarts almost within a second in all the cases – e.g. 20 planes in z-direction, each discretised by a 300×300 grid and 36 azimuthal angles (where only one, the most outer iteration lasts approximately 500 s). The effort to integrate our framework into this application depends on whether the re-allocation of the memory and re-initialisation

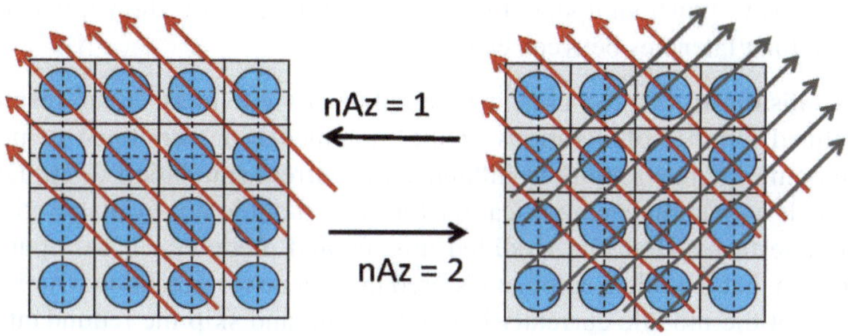

Fig. 5 Experimenting with different numbers of azimuthal angles, small values are given to simplify the picture

of the data is required, and if one wants to re-use the values from the previous iterations [17, 20].

Hierarchical and multilevel approach: It is likely, similar to the heat conduction scenario, that the user wants to accelerate the convergence by starting calculations with lower accuracy (i.e. number of azimuthal angles, see Fig. 5), preserve and re-use some of the values from the previous calculation as an initial guess for the higher accuracy solution. For a conceptually similar algorithm, such as the previously described multilevel approach in the 2D heat conduction simulation, we have seen that our framework has given promising results. The re-initialisation of the data for this, most challenging, scenario is a part of imminent research.

To briefly conclude on this application scenario, the integration of the framework has been straightforward and also not very time consuming. After examining the initial code, deciding which variables to register within the framework, and writing reinitialisation routines, it has taken a few hours to couple the components together and enable visualisation after each iteration. The major effort refers actually to the re-initialisation of variables at the beginning of each "new" computation, i.e. after a user interaction, which is also not a responsibility of the framework itself.

4.3 Test Case 3: Extension of a Problem Solving Environment

Simulation: As mentioned before, SCIRun is a PSE intended for interactive construction, debugging, and steering of large-scale, typically parallel, scientific computations [30]. SCIRun simulations are designed as networks of computational components, i.e. modules connected via input/output ports. This makes it very easy for a programmer to modify a module without affecting others. As SCIRun is already a mature, sophisticated environment for computational steering, nevertheless, our goal is to improve it in a way that real time feedback for extensive time- and memory-consuming simulations becomes possible. Here,

SCIRun needs to finish an update first before new results are shown, which easily can lead to long latencies between cause and effect.

GUI/Visualisation: For the user, it is possible to view intermediate results after a pre-defined number of iterations, while the calculations continue to progress. At some point, he may require to influence the current simulation setup. Different options such as parameter modification for each module are available via corresponding interfaces. Both the modified module and modules whose input data is dependent on that module's output are stored in a queue for execution. Our intention is to interrupt the module currently being executed and skip the redundant cycles, as well as to remove any module previously scheduled for execution from the actual queue.

User Interaction: The concept has been tested on several examples to evaluate the simulation response to the modifications during runtime. These scenarios are: a simulation that facilitates early detection of acute heart ischemia and two defibrillation-like simulations – one on a homogeneous cube and the other on a human torso domain. The challenges of getting an immediate feedback/response of the simulation depend on a few factors – the size of the problem, the choice of the modified parameters within the simulation, etc. The earlier in the execution pipeline the parameter appears, the more modules have to be re-executed, thus, the more challenging it is to provide the real-time response to the user changes. A user can define different discretisation parameters for a FEM computation such as the mesh resolution for all spatial directions. For solving the resulting linear system of equations, different iterative solvers as well as pre-conditioners can be used; one may change tolerances, the maximal number of iterations, levels of accuracy, as well as other numerical or some more simulation-specific parameters. In the created network of modules, typically the most laborious step is the SolveLinearSystem module. Thus, the first challenge is how to interrupt it as soon as any change is made by the user – in particular, the changes done via UI to this module. To achieve this in the algorithm of the linear equation solver, the maximal number of iterations (a user interface variable) is manipulated in the signal handler, so as to be set to some value outside of the domain of the iterator index which interrupts the simulation as described before. The execute function of this module also has to be re-scheduled afterward with the new user-applied settings. However, one has to take care that the previous interrupted execution of the same module is finished in a clean way and that the execute function has to be called anew (in order to trigger re-computation instantly). If one chooses to emit the partial solution after each iteration, executions of several visualisation modules are scheduled after each iteration, which would take additional few seconds after each iteration. This is because after an interrupted iteration the preview of old results has to be cancelled. The execution of all modules, which would happen after SolveLinearSystem, has to be aborted. The scheduler cancels the execution of all the scheduled modules that have not begun yet by making sure an exception is employed. Changing any input field of a module via

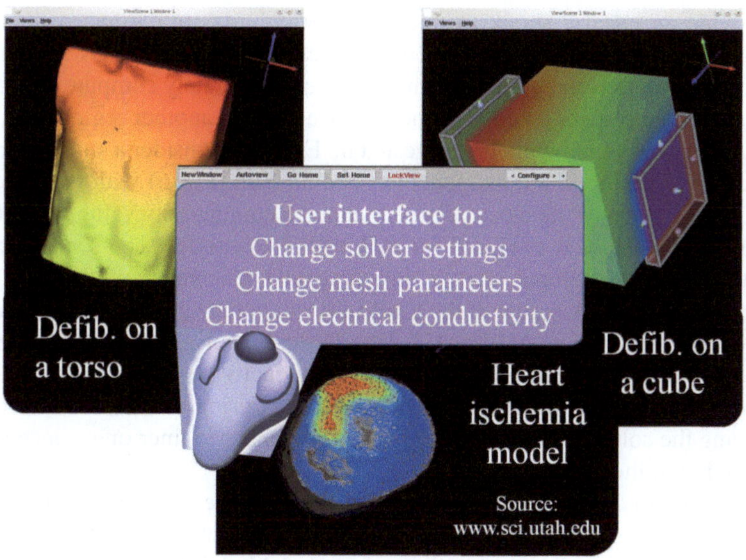

Fig. 6 Illustrated user interfaces for the tested simulation scenario

its UI automatically triggers the re-execution of all the modules following it in the pipeline.

4.3.1 Tool for Early Detection of Heart Ischemia

Myocardial ischemia is characterised by reduced blood supply of the heart muscle, usually due to coronary artery disease. It is the most common cause of death in most Western countries, and a major cause of hospital admissions [28]. By early detection further complications might be prevented. The aim of this application is the generation of a quasi-static volume conductor model of an ischemic heart, based on data from actual experiments [32]. The generation of models of the myocardium is based on MR images/scans of a dog heart. The known values are extracellular cardiac potentials as measured by electrodes on an isolated heart or with inserted needles. The potential difference between the intracellular and extracellular space which is being calculated is not the same for ischemic and healthy cells. A network of modules is constructed within SCIRun to simulate and then render a model of the transmembrane potential of a dog's myocardium in experiments (Fig. 6).

4.3.2 Defibrillation

Defibrillation therapy consists of delivering a dose of electrical energy to the heart with a device that terminates the arrhythmia and allows normal sinus rhythm to be re-established by the body's natural pacemaker. Implantable Cardioverter Defibrillators (ICDs) are relatively common, patient specific, implantable devices that provide an electric shock to treat fatal arrhythmias in cardiac patients [31]. By building a computational model of a patient's body with ICDs and mapping conductivity values over the entire domain, we can accurately compute how activity generated in one region would be remotely measured in another region [36], which is exactly what doctors would be interested in. First, we consider a simulation of the electrical conduction on a homogeneous cube domain (Fig. 6) with two electrodes placed within. Each of the electrodes is assigned a conductivity value. The effect of changing those values is explored for both of the electrodes. The second example helps to determine optimal energy discharge and placement of the ICD in the human torso (Fig. 6). A model of the torso into which ICD geometry is interactively placed is based on patient MRI or CT data. Different solver-related parameters for the resulting system of the linear equations, conductivity values, as well as mesh resolutions for a FEM computation can be applied during runtime. This allows for previewing the solution on a coarser grid and switching to finer ones, once the user is satisfied with the current setting.

For a user to integrate the framework, the major effort has been related to re-triggering the execution of all the needed modules when the user makes a change. This has required good understanding of a used Model-View-Controller pattern. On the other hand, registering the variables which need to be manipulated within the framework to interrupt the execution of the modules of interest, has required negligible amount of time.

4.4 Test Case 4: A Biomedical Application

Another test case is an analysis tool which assists an orthopaedic surgeon to do optimal implant selection and positioning based on prediction of response of a patient-specific bone (femur) to a load that is applied. The tool consists of two coupled components.

Simulation: The first one is a simulation core, where the generated models of femur geometry are based on CT/MRI-data and the computation is done using the Finite Cell Method (FCM). FCM is a variant of high order p-FEM, i.e. convergence is achieved by increasing the polynomial degree p of the Ansatz functions on a fixed mesh instead of decreasing the mesh sizes h as in case of classical h-FEM, with a fictitious domain approach, as proposed in [7]. With this method, models with complicated geometries or multiple material interfaces can be easily handled without an explicit 3D mesh generation. This is especially advantageous for interactive computational steering, where this typically user-interaction intensive step would have to be re-executed for each new configuration.

GUI/Visualisation: The second component is a sophisticated visualisation and user interface platform that allows the intuitive exploration of the bone geometry and its mechanical response to applied loads in the physiological and the postoperative state of an implant-bone in terms of stresses and strains [5, 6]. Thus, after updating the settings – either after insertion/moving an implant, or testing a new position/magnitude of the forces applied to the bone – for each unknown a scalar value, i.e. the so-called von Mises stress norm, can be calculated and the overall result sent to the front end to be visualised.

Some of the challenges in developing such an analysis tool are described in more detail in [5,6,38]. We conveniently had the described simulation and a sophisticated user interface with visualisation module as a starting point. Due to the initial rigid communication pattern between the two components, however, a new setting could be recognised within the simulation only after the results for the previous, outdated, one have been completely calculated and sent to the user. Consequently, the higher polynomial degrees p were used, the dramatically longer became the total time until one could finally perceive the effect of his last change. The integration of our framework then comes into play not only to make the way the data is communicated more suitable for this purpose, but also to enable interrupting the simulation immediately and getting instant feedback ensued by any user interaction.

For the best performance, on the front end, the main thread (in charge of fetching user interaction data and continuous rendering), the second thread (in charge of collecting and sending updates in timely fashion, via non-blocking MPI routines), and the third thread (dedicated for waiting to receive results as soon as these are available), are not synchronised with one another. This way, we tackle the problem of long delays that would occur if one thread is responsible for everything and communication is blocked as long as the thread is busy, which would hinder the user in (smoothly) exploring the effects of his interaction.

On the simulation side, as mentioned before, a variant of FEM is used. Mainstream approaches are

- h-FEM: convergence due to smaller diameters h of elements,
- p-FEM: convergence due to higher polynomial degrees p,
- hp-FEM: combining the aformentioned ones by alternating h and p refinements,
- rp-FEM: a combination of mesh repositioning and p refinements,
- ...

In our case, for the algebraic equations gained by the p-version Finite Element Method describing the behaviour of the femur, iterative solvers such as CG or multi grid could not be efficiently deployed due to the poor condition number of the system. To make most out of the simulation performance potential, a hierarchical concept based on an octree-decomposition of the domain in combination with a nested dissection solver is used [24]. It allows for both the design of sophisticated steerable solvers as well as for advanced parallelisation strategies, both of which are indispensable within interactive applications.

User interaction: By applying a nested dissection solver, the most time consuming step is the recursive assembly of the stiffness matrices, each corresponding to one tree node, traversing the octree bottom up. Again, cyclically-repeating signals are used for frequent checks for updates. If there is an indicator of an upcoming message

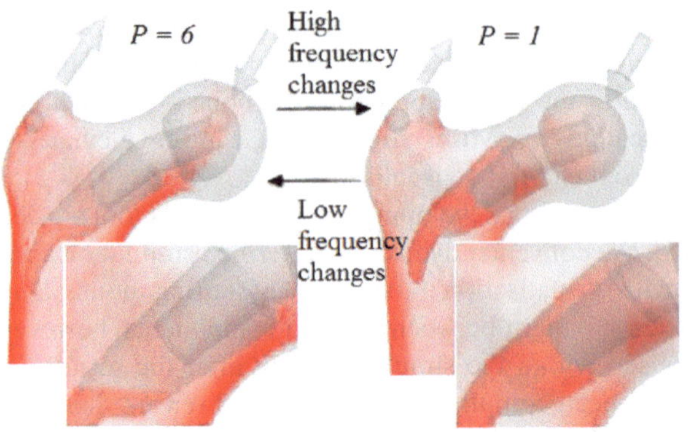

Fig. 7 Direct transition from $p = 6$ to $p = 1$ as soon as the user changes the force's magnitude and direction, inserts an implant or moves it, while gradually increasing from $p = 1 \rightarrow p = 2 \rightarrow p = 4 \rightarrow \ldots$ as soon as the user diminishes or finally stops his interaction. Hence, a qualitative feedback about stress distribution for $p = 1$ or $p = 2$ is received instantly, finer result for $p \geq 4$ on demand

from the user side, this is recognised while processing one of the tree nodes and the simulation variables are set in a way which ensures skipping the rest of them. All the recursive assembly function calls return immediately, and the new data is received in the next step of the interactive computing loop (updating one or more of the leaf nodes). Here, precious time has been saved by skipping all the redundant calculations and, thus, calculating results only for an actual setting. As soon as the whole assembly has been completed without an user interrupt, the result in terms of stresses is sent back to the front end process for visual display. However, there is an unavoidable delay of any visual feedback especially for higher p values, i.e. $p > 4$, in case of the used hardware and the complexity of the geometric model. Namely, the time needed for a (full) new computation is dramatically increasing in case of increasing p. Thus, we profit from a hierarchical approach one more time. The hierarchy exploited in this approach refers to the usage of several different polynomial degrees chosen by the user (Fig. 7). While the user's interplay with the simulation is very intensive, he retrieves immediate feedback concerning the effects of his changes for lower p, being able to see more accurate results (for higher p) as soon as he stops interacting and let one iteration finish. In this case, the computation is gradually switched to higher levels of hierarchy, i.e. from $p = 1$ to $p = 2$ to $p = 4$ and so on. The number of MPI program instances, being executed in parallel for different p can be chosen by the user. A detailed communication schemes can be found in [16, 18].

To get several updates per second even for higher p values, one has to employ sophisticated parallelisation strategies. Custom decomposition techniques (i.e. recursive bisection) in this scenario, as in case of long structures such as a femur, typically hinder the efficient exploitation of the underlying computing power as this leads to improper load distributions due to large separators within the nested dissection approach. Thus, our next goal has been the development of an efficient load balancing strategy for the existing structural simulation of the bone stresses.

Task scheduling strategies typically involve a trade-off between the uniform work load distribution among all processors, as well as keeping both the communication and optimisation costs minimal. For hierarchically organised tasks with bottom-up dependencies, such as in our generated octree structure, the number of processors participating in the computation decreases by a factor of eight in each level, similar to the problem posed by Minsky for the parallel summation of $2N$ numbers with N processors in a binary tree [19].

In interactive applications which assume the aforementioned frequent updates from user's side, those rapid changes within the simulation and tasks' state favour static in comparison to dynamic load balancing strategies. It would also have to be taken into consideration that certain modifications performed by a user may involve major changes of the computational model. In this case, for repeatedly achieving the optimal amount of work being assigned to each process for each new user update, the overhead-prone scheduling step has to be executed each time. Therefore, an efficient, nevertheless simple to compute scheduling optimisation approach is needed.

Since the scheduling problem can be solved by polynomial-depth backtrack search, thus, is *NP* complete for most of its variants, efficient heuristics have to be devised. In our case, the sizes of the tasks, as well as the dependencies among them (given by the octree structure responsible for the order of the nested dissection advance) have to be considered. By making decisions, we consider (1) the level of the task dependency in the tree hierarchy where children nodes have to be processed before their parent nodes; (2) among equal tasks (i.e. of the same dependency level) we distinguish between different levels in the tree hierarchy, calling this property the processing order. If the depth of the tree is H, tasks from level M in the tree hierarchy have the processing order of $H - M - 1$. Then we form lists of priorities, based on these two criteria, since tasks inside very long branches of the tree with an estimated bigger load should be given a higher priority. Additionally, we resort to a so-called max-min order, making sure that big tasks, in terms of their estimated number of floating-point operations, are the first ones assigned to the processors. We also split a single task among several processors when mapping tasks to processors, based on the comparison of a task's estimated work with a pre-defined 'unit' task. This way, arrays of tasks, so-called 'phases', are formed, each phase consisting of as many generated tasks as there are computing resources. Namely, taken from the priority lists, tasks are assigned to phases in round-robin manner. The results are illustrated in Fig. 8.

Those phases refer to the mapping which will be done during runtime of the simulation. When the tasks are statically assigned to the processors, all of them execute the required computations, communicating the data when needed and also

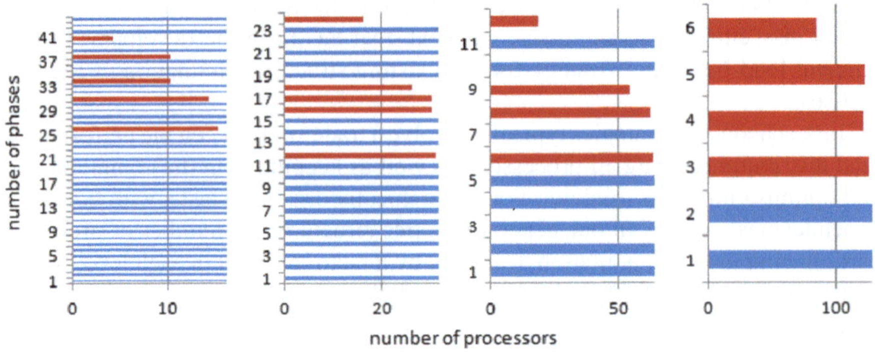

Fig. 8 *Vertical axes* describe the so-called "phases" and the *horizontal axes* the number of processors involved in the particular phase. One "phase" involves actually the processors to which a task is assigned at that point. Having the capacity of each phase as *full* as possible is achieved, i.e. all processors are busy with the approximately equal amount of work throughout the solver execution

taking care that the communication delays due to the MPI internal decisions are avoided, as elaborated more in [19].

Satisfactory speedup is achieved for different polynomial degrees p within the FCM computation, where higher polynomial degrees correspond to more unknowns. Tests are being done currently for a larger number of distributed memory computational resources. According to the tendency observed for up to seven processors so far, engagement of larger numbers of processes would result in the desired rate of at least several updates per second (i.e. 1–10 Hz) for the calculated bone stresses even for $p = 4$ or $p = 6$.

Referring back to the existing environment, without the integration of the developed distributed parallel solver, the major effort that has been invested in creating the new communication pattern to support the described hierarchical approach was in the order of several working day. Anyhow, the functionality for interrupting the computation to do checks for updates, thus, start a computation anew if needed has been quick and straightforward.

5 Results and Conclusions

Finally, after discussing the achievements concerning interaction for each application scenario in the previous section, here results in terms of execution time overhead after integrating the framework in different scenarios are to be presented, as well as the coding effort to be invested when integrating the framework into an existing application code. Furthermore, conclusions concerning the proposed

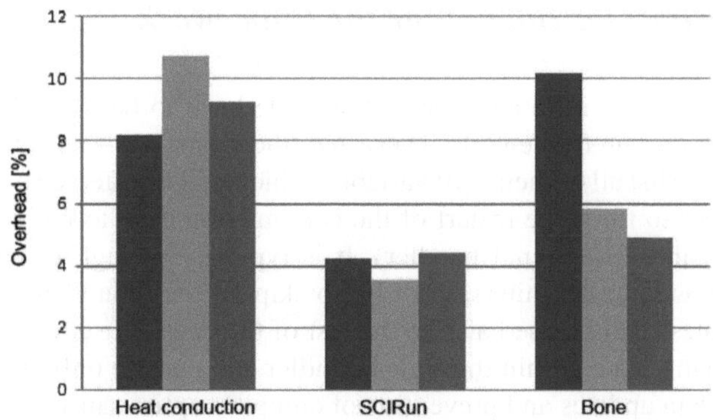

Fig. 9 Performance measurements: overhead of the framework (expressed in terms of additional execution time) for alarm set to 1 ms – heat conduction simulation (300 × 300 grid), executed on 1, 2, and 4 cores (left to right); SCIRun PSE, heart ischemia example using CG, BCG, and MINRES solver (left to right); biomedical application ($p = 4$), executed on 1, 2, and 4 cores (left to right)

hierarchical approaches are made and possible ideas for further extension of the framework are discussed.

5.1 *Overhead of the Framework*

For the heat conduction application scenario, the integration of our framework resulted in not more than 5–10 % overhead in the execution time. Tests have been done also for the same problem with a message-passing-based parallel Jacobi solver. Not even in case when user interaction was invoked in 5-ms intervals (which is far more frequent than it typically occurs in practice) any significant effect of the interrupts on the overall execution time (less than 10 %) was to be observed.

Performance evaluation of the biomedical test scenario, where the simulation is executed on a multi-core architecture and connected to a visualisation front end via a network, still proved that the overhead caused by the framework itself is not significant (up to 11.7 %).

We have also tested the different simulation scenarios from SCIRun. The measurements have been made for different update intervals, namely, 5, 2, or 1 ms for different solvers of linear systems of equations. In one of the test case scenarios, for the shortest interval (i.e. 1 ms), the overhead caused by the framework was up to 15 %. However, by making the intervals longer (2 or 5 ms, e.g.), the overhead was reduced to 5 and 3 %, resp. When increasing the interval up to 5 ms (and beyond), an end-user does not observe the difference in terms of simulation response. Hence, it is always recommendable to experiment with different intervals for a specific simulation.

Some of the measurements are illustrated in Fig. 9 for comparison.

5.2 User Effort for Integrating the Framework

A few modifications within any application code have to be made by the user in order to integrate our framework. These modifications are – as intended – only minor, hence, we list all of them. All variables which will be affected by the interrupt handler in order to force the restart of the computation have to be declared global (to become visible in a signal handler). It is typically enough to have only few of them, such as loop delimiters, in order to skip all the redundant computations. If these variables shall be used also in the rest of the code, a user can rename those he wants to manipulate within the signal handler and declare only those as global. Atomicity of data updates and prevention of compiler optimisations – which would lead to incorrect value references – have to be ensured. The integrity of each user-defined 'atomic' sequence of instructions in the simulation code has to be provided. The calls to the appropriate send and receive functions which are interface to our framework have to be included in the appropriate places in the code. The user himself should provide the correct interpretation of the data (in the receive buffers of both simulation and visualisation components). Finally, he has to enable the regular checks for updates by including appropriate functions which will examine and change the default signal (interrupt) action, specifying the time interval in which the checks of the simulation process(es) are made, as shown in the following pseudo code example.

```
% Function to override the default signal action
begin func My_sig_action ()
  if update_available then
    receive update
    manipulate simulation specific variables
  fi
end

% Declare simulation specific variables to be
% global, atomic, and volatile
begin func main ()
  Set_sig_action (My_sig_action)
  Set_interrupt_interval (time_slot)
end
```

5.3 Hierarchical Approaches

As one may also conclude, no matter how generic our basic idea is, when applying it to the wide diversity of applications, the user himself has to be involved in making certain decisions. For example, in our first test case, he has to specify the number of grids which he would like to use together with their resolutions. This information might be based on his previous experience, i.e. at which resolution the problem can be solved within less than a second (for choosing the coarsest grid), etc. The hierarchical approaches used so far should not be the limitation for future test cases. In addition to recursively coarsening the grid, or increasing the resolution of other simulation-specific discretisations such as the number of azimuthal angles in AGENT, or increasing the polynomial degree p in the biomedical example, one may analogously profit from his or her own simulation-specific hierarchical structures. Any user of the framework can, if needed, easily adopt it to his individual requirements.

5.4 Outlook

In the future, we would like to tackle the computational expensive scenarios with massively parallel simulations. In efforts to interrupt one thread per process, a trade-off between ensuring a minimal number of checks per process and allowing for receiving the data promptly is to be faced. Thus, an optimal interval between the interrupts on different levels of the communication hierarchy is going to be estimated. In addition, a possibility of distributing the tasks among several user processes, each in charge of a certain group of simulation processes will be examined to avoid typical master-slave bottlenecks. Furthermore, we would like to explore techniques for the fast transfer of (distributed) simulation results between front and back end, especially in case of huge data sets, needed for an interactive visualisation.

Acknowledgements The overall work has been financially supported by the Munich Centre of Advanced Computing (MAC) and the International Graduate School of Science and Engineering (IGSSE) at Technische Universität München and we would like to gratefully acknowledge that. The work related to SCIRun PSE was made possible in part by software from the NIH/NIGMS Center for Integrative Biomedical Computing, 2P41 RR0112553-12. It was accomplished in winter 2011/12 during a 3-month research visit of Jovana Knežević to the Scientific Computing and Imaging (SCI) Institute, University of Utah. She would like to express her appreciation and gratitude to Prof. Chris Johnson for inviting her and all the researchers for fruitful discussions. Furthermore, she would like to thank Hermilo Hernández and Tatjana Jevremović at Nuclear Engineering Program, University of Utah, and Thomas Fogal from SCI Institute, in collaboration with whom the work on the AGENT project was done.

References

1. Baolai G.: Fortran signal handling. In: Proceedings of the SHARCNET Workshop, Hamilton (2008)
2. Borrmann, A., Wenisch, P., van Treeck, C.: Collaborative HVAC design using interactive fluid simulations: a geometry-focused platform. In: Proceedings of the 12th International Conference of Concurrent Engineering, Fort Worth (2005)

3. Brooke, J., Coveny, P., Harting, J., Jha, S., Pickles, S., Pinning, R., Porter, A.: Computational steering in RealityGrid. In: Proceedings of the UK e-Science All Hands Meeting, Nottingham (2003)
4. de St. Germain, J., McCorquodale, J., Parker, S., Johnson, C.: Uintah: a massively parallel problem solving environment. In: Proceedings of the 9th International Symposium on High-Performance Distributed Computing, Pittsburgh, pp. 33–41 (2000)
5. Dick, C., Georgii, R., Burgkart, R., Westermann, R.: Computational steering for patient-specific implant planning in orthopedics. In: Proceedings of the 1st Eurographics Conference on Visual Computing for Biomedicine, Delft, pp. 83–92 (2008)
6. Dick, C., Georgii, R., Burgkart, R., Westermann, R.: Stress tensor field visualisation for implant planning in orthopedics. IEEE Trans. Vis. Comput. Graph. **15**(6), 1399–1406 (2009)
7. Düster, A., Parvizian, J., Yang, Z., Rank, E.: The finite cell method for three-dimensional problems of solid mechanics. Comput. Methods Appl. Mech. Eng. **197**(45–48), 3768–3782 (2008)
8. Fogal, T., Krüger, J.: Tuvok, an architecture for large scale volume rendering. In: Proceedings of the 15th International Workshop on Vision, Modelling, and Visualization, Siegen (2010)
9. Geist, G., Kohl, J., Papadopoulos, P.: CUMULVS: Providing fault-tolerance, visualization and steering of parallel applications. Int. J. High Perform. Comput. Appl. **11**(3), 224–236 (1997)
10. http://www.realitygrid.org. RealityGrid
11. http://www.wxWidgets.org. WxWidgets Library
12. Jenz, D., Bernreuther, M.: The computational steering framework steereo. In: Proceedings of the PARA 2010 Conference: State of the Art in Scientific and Parallel Computing, Iceland (2010)
13. Knezevic, J., Mundani, R.P.: Interactive computing for engineering applications. In: Proceedings of the 22nd Forum Bauinformatik, Karlsruhe, pp. 137–144 (2010)
14. Knezevic, J., Frisch, J., Mundani, R.P., Rank, E.: Interactive computing framework for engineering applications. In: Proceedings of the INTERCOMP 2011, International Conference on Computer Science and Applied Computing, Vienna (2011)
15. Knezevic, J., Frisch, J., Mundani, R.P., Rank, E.: Interactive computing framework for engineering applications. J. Comput. Sci. **7**(5), 591–599 (2011)
16. Knezevic, J., Mundani, R.P., Rank, E.: Interactive computing in preoperative planning of joint replacement. In: Proceedings of the International Conference on Modeling, Simulation and Control, Singapore, pp. 86–91 (2011)
17. Knezevic, J., Hernandez, H., Fogal, T., Jevremovic, T.: Visual simulation steering for a 3d neutron transport agent code system. In: Proceedings of the INREC International Nuclear and Renewable Energy Conference, Amman (2012)
18. Knezevic, J., Mundani, R.P., Rank, E.: Interactive computing – virtual planning of hip-joint surgeries with real-time structure simulations. J. Model. Optim. **1**(4), 308–313 (2012)
19. Knezevic, J., Mundani, R.P., Rank, E.: Schedule optimisation for interactive parallel structure simulations. In: Proceedings of the PARA 2012 Workshop on State-of-the-Art in Scientific and Parallel Computing, Helsinki (2012)
20. Knezevic, J., Mundani, R.P., Rank, E., Hernandez, H., Jevremovic, T., Fogal, T.: Interactive computing in numerical modelling of particle transfer methods. In: Proceedings of the IADIS International Conference on Theory and Practice in Modern Computing, Lisbon (2012)
21. Lee, H., Retzke, K., Peng, Y., Jevremovic, T., Hursin, M.: AGENT code: Open-architecture analysis and configuration of research reactors—neutron transport modeling with numerical examples. In: Proceedings of the PHYSOR—The Physics of Fuel Cycles and Advanced Nuclear Systems: Global Developments, Chicago (2004)
22. Mulder, J., van Wijk, J.: Logging in a computational steering environment. In: Proceedings of the 6th Eurographics Workshop on Visualization in Scientific Computing '95, Sardegna, pp. 118–125 (1995)
23. Mulder, J., van Wijk, J., van Liere, R.: A survey of computational steering environments. Future Gener. Comput. Syst. **15**, 119–129 (1999)

24. Mundani, R.P., Bungartz, H.J., Rank, E., Niggl, A., Romberg, R.: Extending the p-version of finite elements by an octree-based hierarchy. In: Domain Decomposition Methods in Science and Engineering XVI, pp. 699–706. Springer, Berlin/Heidelberg (2007)
25. Nicolas, R., Esnard, A., Coulaud, O.: Toward a computational steering environment for legacy coupled simulations. In: Proceedings of the 6th International Symposium on Parallel and Distributed Computing, Hagenberg (2007)
26. Parker, S., Miller, M., Hansen, C., Johnson, C.: An integrated problem solving environment: The SCIRun computational steering system. In: Hawaii International Conference on System Sciences, Kohala Coast, pp. 147–156. IEEE (1998)
27. Pickles, S., Haines, R., Pinning, R., Porter, A.: Computational steering in RealityGrid. In: Proceedings of the UK e-Science All Hands Meeting, Nottingham (2004)
28. Podrid, P., Myerburg, R.: Epidemiology and stratification of risk for sudden cardiac death. Clin. Cardiol. **28**, I3–11 (2005)
29. Rycerz, K., Bubak, M., Sloot, P., Getov, V.: Problem solving environment for distributed interactive applications. In: Proceedings of the CoreGRID Integration Workshop, Krakow, pp. 129–140 (2006)
30. Shepherd, J., Johnson, C.: Hexahedral mesh generation for biomedical models in SCIRun. Eng. Comput. **25**(1), 97–114 (2009)
31. Steffen, M., Tate, J., Stinstra, J.: Defibrillation tutorial. Scientific Computing & Imaging Institute (2012). http://www.sci.utah.edu/devbuilds/scirun_docs/DefibrillationTutorial.pdf
32. Stinstra, J., Swenson, D.: Ischemia model tutorial. Scientific Computing & Imaging Institute (2012). http://www.sci.utah.edu/devbuilds/scirun_docs/IschemiaModelTutorial.pdf
33. van Liere, R., van Wijk, J.: CSE: A modular architecture for computational steering. In: Virtual Environments and Scientific Visualization '96, pp. 257–266. Springer, Vienna (1996)
34. van Treeck, C., Wenisch, P., Borrmann, A., Pfaffinger, M., Egger, M.: Utilizing high performance supercomputing facilities for interactive thermal comfort assessment. In: Proceedings of the 10th International IBPSA Conference Building Simulation, Beijing (2007)
35. Vetter, J., Schwan, K.: High performance computational steering of physical simulations. In: Proceedings of the 11th International Parallel Processing Symposium, Geneva (1997)
36. Weinstein, D., Parker, S., Simpson, J., Zimmerman, K., Jones, M.: Visualization in the SCIRun problem-solving environment. In: Visualization Handbook, pp. 615–632. Elsevier, Burlington (2005)
37. Wenisch, P., van Treeck, C., Rank, E.: Interactive indoor air flow analysis using high performance computing and virtual reality techniques. In: Proceedings of Roomvent, Coimbra (2004)
38. Yang, Z., Dick, C., Düster, A., Ruess, M., Westermann, R., Rank, E.: Finite cell method with fast integration – an efficient and accurate analysis method for CT/MRI derived models. In: Proceedings of ECCM, Paris (2010)

A Framework for the Interactive Handling of High-Dimensional Simulation Data in Complex Geometries

Amal Benzina, Gerrit Buse, Daniel Butnaru, Alin Murarasu, Marc Treib, Vasco Varduhn, and Ralf-Peter Mundani

Abstract Flow simulations around building infrastructure models involve large scale complex geometries, which when discretized in adequate detail entail high computational cost. Moreover, tasks such as simulation insight by steering or optimization require many such costly simulations. In this paper, we illustrate the whole pipeline of an integrated solution for interactive computational steering, developed for complex flow simulation scenarios that depend on a moderate number of both geometric and physical parameters. A mesh generator takes building information model input data and outputs a valid cartesian discretization. A sparse-grids-based surrogate model—a less costly substitute for the parameterized simulation—uses precomputed data to deliver approximated simulation results at interactive rates. Furthermore, a distributed multi-display visualization environment shows building infrastructure together with flow data. The focus is set on scalability and intuitive user interaction.

Keywords Building infrastructure models • Computational steering • Surrogate models • Multi-display visualization • 3D gesture-based interaction

1 Introduction

Today, the need for energy efficient buildings and optimized constructions is high and will increase dramatically in the future, a trend forced by ongoing shortage of fossil resources and increasing environmental awareness of society. This also has a share in making numerical simulations an indispensable tool in various fields when physical phenomena or other complex processes need to be analyzed and

A. Benzina (✉) · G. Buse · D. Butnaru · A. Murarasu · M. Treib · V. Varduhn · R.-P. Mundani
Technische Universität München, Munich, Germany
e-mail: benzina@in.tum.de

M. Bader et al. (eds.), *Advanced Computing*, Lecture Notes in Computational Science and Engineering 93, DOI 10.1007/978-3-642-38762-3_10,

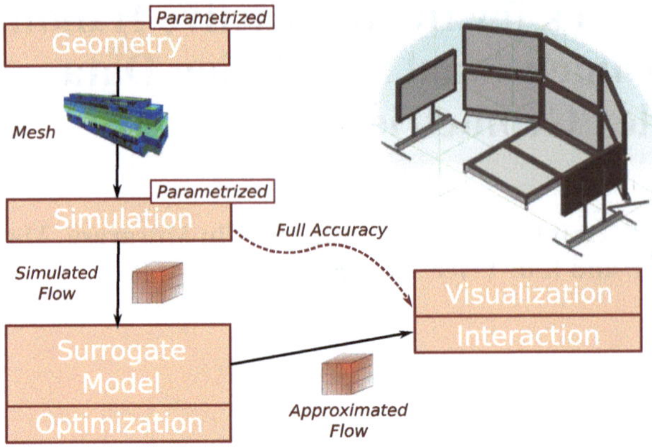

Fig. 1 The image sketches our integrated computational steering solution which features interactive and intuitive visual exploration of parametrized geometries and simulations. An optimized surrogate model is used in conjunction with the actual simulation to ensure fast delivery of approximated simulation results to the multi-display visualization component

understood. Often it is even necessary to run simulations repeatedly with slightly adjusted boundary conditions, initial conditions, or even geometry.

While in fields such as automotive and aerospace engineering the consistent application of numerical simulations in the development process has been state of the art for a long time already, this is not the case yet for civil engineering. Reasons for this situation range from engineers still sticking to paper-based construction plans to the simple fact that the integration of simulation technology into the process still lacks support.

Yet, the advance of building information model (BIM) data during the last decade now motivates a building performance analysis framework for construction and built infrastructure. BIMs provide a fully-detailed product model for constructions, including exact geometric representation and auxiliary information such as material parameters or measured information. Parameter adjustment is also supported in BIMs, which is crucial to performance evaluation of buildings, such as an indoor temperature analysis on a hot summer day depending on different window and door opening angles and the intensity of the air conditioning system. The consideration of all such factors leads to a large design space which can be explored by means of numerical simulations. However, the size of this space and the computational cost of computational fluid dynamics (CFD) simulations call for an elaborate approach to interactive steering.

In this work, an integrated solution for computational steering of parameterized flow simulations in complex, large-scale geometries is presented. Figure 1 gives an overview of the system components and their interaction. The starting point in the pipeline is the component described in Sect. 2 capable of maintaining high-detail product model data. It also supports several data pre-processing steps for the actual simulation (described in detail in Sect. 3), such as mesh generation, domain decomposition and the application of boundary conditions.

Next, our implementation of a surrogate model as a substitute for simulations is introduced and analyzed for performance. The surrogate model is based on the *Sparse Grid Technique* and described in detail in Sect. 4. At the core of the model, simple vector operations need to be optimized to ensure the performance needed in interactive settings. Various platform-related considerations and benchmark results for modern NUMA, i.e. Non-Uniform Memory Access, platforms are found in Sect. 5.

Finally, Sects. 6 and 7 focus on a semi-immersive visualization with integrated support for 3D interaction with the data. The investigated environment consists of eight 3D monitors with various interaction devices such as motion sensing input devices and depth cameras. Special focus is set to the challenges of fast interaction with multi-dimensional data and of an intuitive user interface that enhances the exploration process.

2 Data

Dealing with data from constructions and built infrastructure, we implement Industry Foundation Classes (IFC) [6]. On the one hand, this involves the complete product model description for construction data, i.e. the fully detailed geometry specification together with auxiliary data such as material parameters and measured information. On the other hand, IFC are the de-facto industry standard for exchanging construction data and therefore open the door for incorporating data from almost all engineering sources.

This functionality is integrated into a framework for fast access to large sets of construction and built infrastructure data, which meets the requirements imposed by CFD simulations targeting high accuracy.

Several level-of-detail metaphors have been developed, based on stepwise coarsening of the fully-detailed data. Approximating complex geometric entities with simpler primitives and sorting out irrelevant information thus allows for efficient data delivery at various resolutions.

In order to exclude unimportant details from further processing, an algorithm has been developed to recognize those parts of a construction that are visible from the outside. In a pre-processing step, each triangle is rendered to an image using a unique color code. Then, all triangles visible from the respective point of view are identified by means of their color code appearing in the image. Figure 2 shows the color coding applied to the TUM building.

By rotating around the z and y axes and combining the visibility information from all steps, the set of triangles forming the outer hull is identified. Typically, a reduction of over 90 % in the number of triangles can be achieved, as the outer hull usually consists of less than 10 % of the whole construction's triangles (see Fig. 3). For a detailed description of the algorithm, the reader is referred to [16].

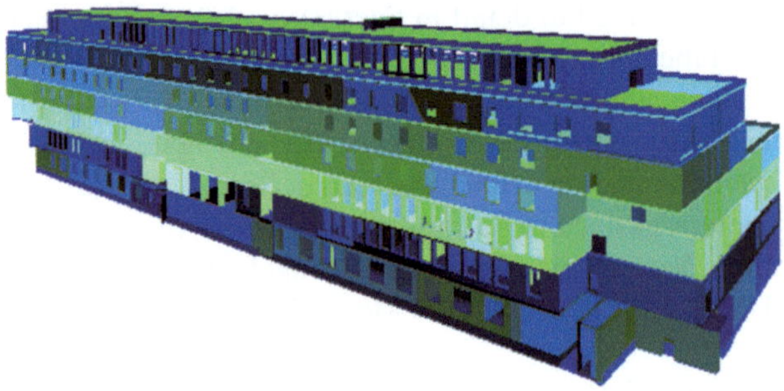

Fig. 2 The triangle color coding applied to the TUM building. The ID of each triangle is represented as a 24 bit RGB color, therefore up to 16 million triangles can be identified

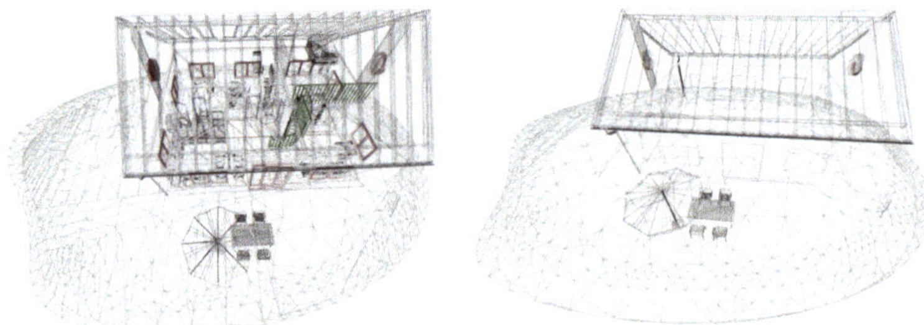

Fig. 3 In the *left picture*, the full product model is visualized, whereas in the *right picture* the visualization is reduced to the triangles identified by the outer hull algorithm, which are approximately 10 %

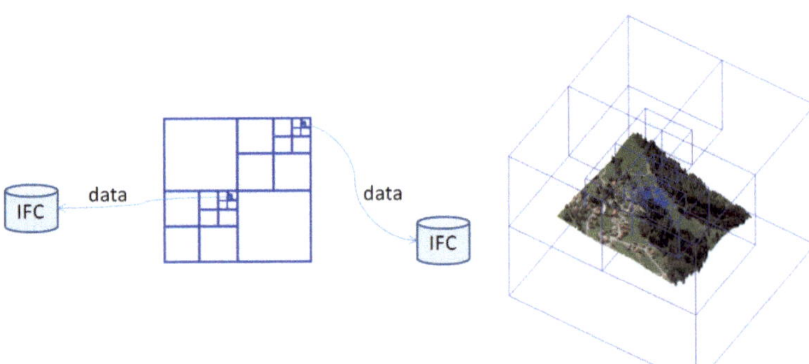

Fig. 4 In the *left picture*, the quadtree is sketched, which is built up on the locations of the product models and only points to their storage information. The *right picture* indicates the embedding of the stored construction and terrain data into the hierarchical data structure

As it has already been pointed out, fast access to the underlying data is mandatory for performing the analysis presented in this work. To this end, a two-layer access strategy incorporating hierarchical data structures has been developed. As shown in Fig. 4, a static octree representation of all constructions in the computational

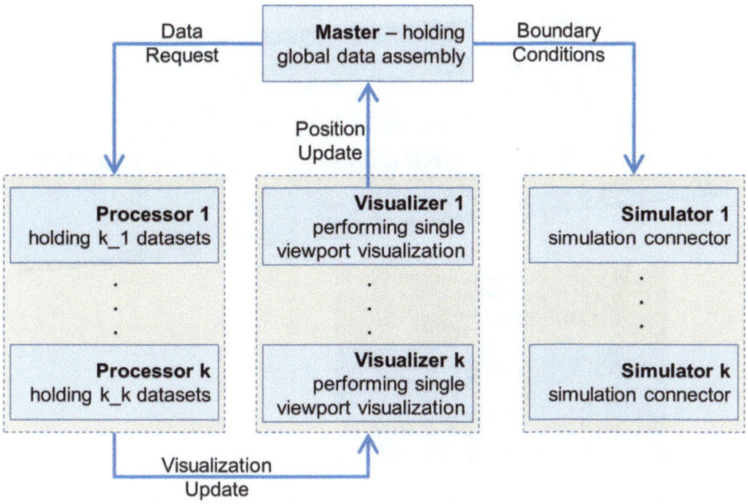

Fig. 5 Orchestration of the different process types with the flow of data, i.e. communication, in the parallel organization of the framework

domain is built up, giving global access to all buildings' bounding boxes and storage meta-data. This first-level octree thus ensures global location awareness and orchestrates a set of second-level octrees, each of which holds the data for a single construction. The constructions themselves are only processed on demand whenever local operations are to be performed on the data. For a detailed description, the reader is referred to [15].

Based on the above data access concept, an optimized toolbox for further tasks such as voxel data generation (e.g. for numerical simulations) or visualization has been compiled as illustrated in Fig. 5. The first-level octree offers convenient access to the full set of meta-data and thus simplifies the task of implementing a global domain decomposition strategy. The hierarchical nature of the model further helps splitting the data into large parts that are independent with respect to the task at hand. For further reading please refer to [16].

3 Simulation

Computational fluid dynamics (CFD) simulations are especially challenging and costly, which makes them a perfect fit to try and apply our computational steering approach. The goal in the scenario of our choice is to examine the flow around and through the main building of Technische Universität München (TUM) with a special focus set on the influence of several varying parameters. The building's highly detailed model (more than 130k triangles) is obtained by applying the tools from Sect. 2 in order to extract the geometric specification from an Industry Foundation

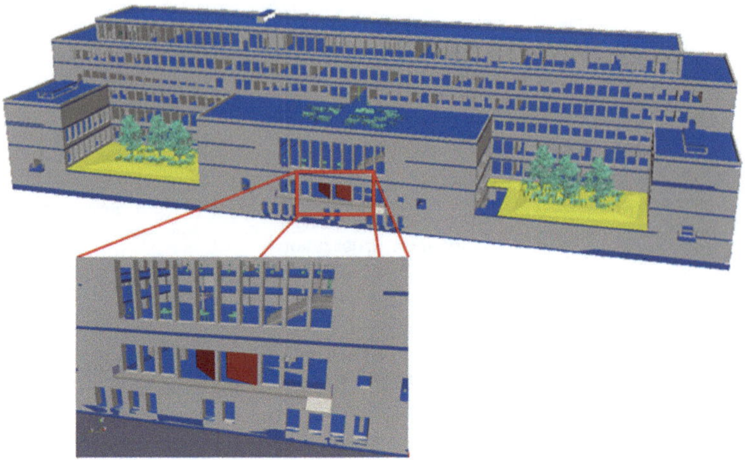

Fig. 6 As a complex geometry we consider the main building of TUM. Within the building we select two doors (highlighted in *red*) whose opening angles are geometric parameters that can be varied independently. Each angle is a dimension in the simulation space

Classes (IFC) [6] product model. The three parameters in the setting are the opening angles of two large doors, and the inlet velocity of the fluid modeled as domain boundary condition. Figure 6 provides a visual clue about the location of the doors, the flow is directed towards the front of the building.

Both doors are independently adjusted with opening angles in the range [0°; 90°], and the incoming flow assumes values in [5; 15 m/s] at the boundary. The computational domain has a physical size of 48 × 12 × 12 m and is discretized by a regular, equidistant mesh of 1,024 × 256 × 256 grid points.

As further explained in Sect. 4, the total number of simulation runs performed amounts to 111, where each run is configured with a different parameter combination.

Instead of implementing our own CFD solver, we conceived the pre-processing toolkit presented in Sect. 2 such that a binding to existing simulation software would be possible. In the simulation step of the pipeline we therefore take advantage of the capabilities of the *potentialFoam* solver of the open source library *OpenFOAM* v. 2.1.1 [11]. At the beginning of each simulation run it provides mass conserving initial values and is configured to solve the pressure equation once, before proceeding with the solution of the incompressible 3D Navier-Stokes equations. In order to complete this task in a manageable amount of time, the software has been ported to the Shaheen supercomputer at KAUST [7] (BlueGene/P architecture), which is equipped with approximately 65k cores and 1 GB main memory per core. While load distribution has been achieved through a blocked, checkerboard-like domain decomposition scheme, every simulation run still consumed approximately 230–300 core hours for the initialization of the flow field.

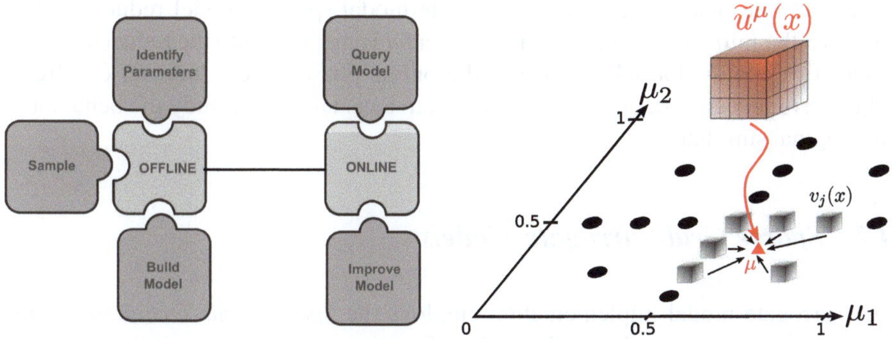

Fig. 7 *Left*: phases of a surrogate model construction and use. *Right*: The sparse grid sampling is given by the sparse grid discretization (*black circles*). Approximated snapshots for a requested parameter combination (*red triangle*) are constructed as a linear combination of affected bases (*small cubes*)

4 Computational Steering by Means of High-Dimensional Interpolation

The design of efficient buildings and optimized constructions is the result of an interplay between various criteria ranging from material constraints and regulations to thermal efficiency or acoustic comfort. Many of these criteria can be mapped to parameters in a numerical simulation tool, which can then be used to identify optimal values or to obtain insight regarding the connection between parameters and simulation outcome. Both tasks, optimization and simulation insight, require running a potentially large number of simulations. However, an increasing number of parameters combined with high computational cost of single simulations (cf. Sect. 3) make direct simulation techniques unfeasible and raise questions about alternatives.

4.1 Surrogate Models for Computational Steering

As detailed in [3], a parameterized simulation can be considered as a high-dimensional function $u : \Omega \times \mathscr{P} \to \Gamma$, which maps the spatial domain Ω and the parameter space $\mathscr{P}$ onto the space of simulation results Γ (e.g. the cartesian product of the simulation's primary quantities). Surrogate models substitute the function u for a simplified form $\tilde{u} : \Omega \times \mathscr{P} \to \Gamma$, from which approximated simulation results can be retrieved in much shorter time.

Figure 7a illustrates the main parts of the surrogate model approach. The construction of the model is done *offline* and starts by choosing the parameters and their ranges of interest. The (normalized) resulting parameter space $\mathscr{P}$ is then sampled, and with each sample corresponding to one parameter combination, a full simulation needs to be computed per sample. We will denote the set of sampling points $\mathscr{P}_s \subset \mathscr{P}$.

The offline phase ends with a surrogate-model-specific model reduction step, in which the initial high-dimensional problem u is mapped onto the reduced form $\tilde{u}$. Interactive exploration takes place in the online phase, where the user can freely adjust parameters, and, guided by an interactive visualization, study the behavior of the original simulation.

4.2 Sparse Grid Surrogate Models

Our surrogate model implementation employs the *Sparse Grid Technique*, taking advantage of its extremely good cost-benefit ratio regarding numerical approximation in higher dimensions. For smooth functions, sparse grids manage to maintain a good approximation quality while keeping the set of sampling points $\mathscr{P}_s$ small (see [1]). The smoothness of our simulation function is not a priori known, but adaptive sparse grids can relax the smoothness requirements (see [12]) by investing sampling points based on the local gradient information. In the context of surrogate models this means that during the offline phase only few simulations need to be performed, one for each $\mu_j \in \mathscr{P}_s,\ j \in \{1,\dots,M\}$. In the following we will refer to the results of a single simulation as *snapshot*, i.e. the component function

$$u(\mathbf{x},\mu) = u^{\mu}(\mathbf{x}),\ \mathbf{x} \in \Omega \tag{1}$$

is a snapshot defined by parameter combination $\mu \in \mathscr{P}$.

The reduced form $\tilde{u}$ is obtained by building a multi-dimensional sparse grid interpolant from a particular set of snapshots. When evaluated at position $\mu \in \mathscr{P}$, $\tilde{u}$ returns a data vector for the approximated snapshot $\tilde{u}^{\mu}(\mathbf{x})$ resulting from interpolation between the precomputed snapshots $u^{\mu_j}(\mathbf{x}),\ \mu_j \in \mathscr{P}_s$. Technically, the evaluation step is a linear combination

$$\tilde{u}^{\mu} = \sum_{j=1}^{M} \hat{u}_j\ \phi_j(\mu) \tag{2}$$

where the $\phi_j,\ j \in \{1,\dots,M\}$ denote the hierarchical sparse grid basis functions. The $\hat{u}_j$ are the same as the snapshots u^{μ_j}, only with respect to the hierarchical sparse grid basis. For more information regarding sparse grid interpolation we refer to [1]. Equation 2 is the main motivation for employing sparse grids in our setting as it can be mapped quite efficiently onto modern hardware (see Sect. 5).

4.3 Surrogate Model for Building Infrastructure Simulations

CFD simulations around building infrastructure are particularly computationally expensive due to the high resolution required to capture adequate detail, but also

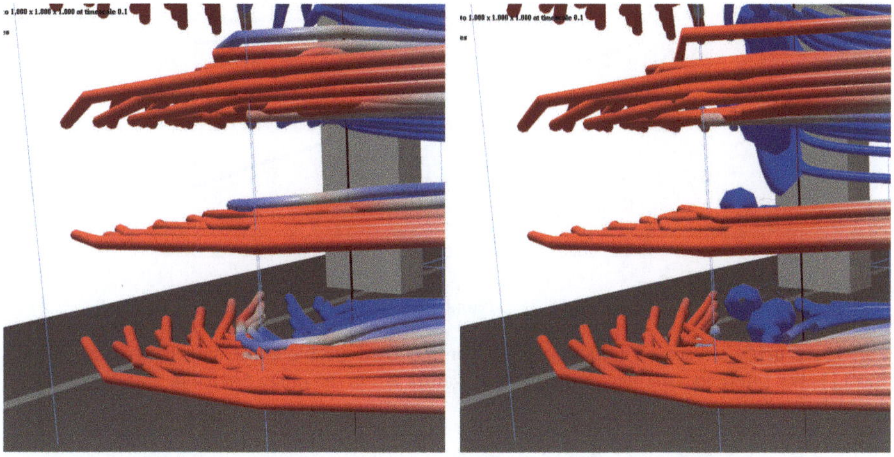

(a) $|\mathscr{P}_s| = 111$ (b) Reference

Fig. 8 (**a**) Interpolated flow (zoom in close to the doors) with door angles 32.88° and 83.28° and inflow velocity 12.5 m/s. (**b**) Reference solution from the full simulation at the same parameter combination

because of the large computational domain (up to kilometers). The setting described in Sect. 3 does not bear such large spatial expansion, yet interactive computational steering is still only possible by means of surrogate models.

As initial grid for the surrogate model, we chose a sparse grid of level $l = 3$ to sample the normalized three-dimensional parameter space $\mathscr{P}$. This entails running 31 simulations in the offline phase, each computing a data block of size $[N_x \times N_y \times N_z \times N_{DOF}] = [1{,}024 \times 256 \times 256 \times 4]$, where N_{DOF} is the number of degrees of freedom per simulation vertex (here: 4 components from velocity field U and pressure P). For the base grid the total output thus amounts to 31 (simulations) $\times$ 1 GB = 31 GB.

The sparse grid surrogate has been shown to provide reasonable accuracy (1 % relative L2 error) for various simulation codes (see [4]). We discuss next the observed accuracy for the BIM simulation. From a visual point of view, Fig. 8 shows interpolated results for a parameter combinations which match the expected flow behavior. We consider $\mathscr{P}_t$ as a set of test points where the interpolation error is to be measured and construct it by sampling 125 points in a uniform $5 \times 5 \times 5$ grid from the parameter interval $[0.12, 0.92] \times [0.12, 0.92] \times [0.12, 0.92]$ (normed parameter ranges). The surrogate with $|\mathscr{P}_s| = 31$ delivers over $\mathscr{P}_t$ a maximum absolute error of $2\text{e}-04$ and an average error of $1\text{e}-04$ (see Fig. 9).

The approximation $\tilde{u}$ can be improved by investing more computational time in the offline phase and thus extending the sparse grid in an adaptive manner. We perform two such extensions by first refining the $|\mathscr{P}_s| = 31$ model which leads to a surrogate model with $|\mathscr{P}_s| = 87$. A second refinement increases the model size to $|\mathscr{P}_s| = 111$. As expected, we observe an increase in accuracy of the two models over the initial $|\mathscr{P}_s| = 31$ (see Fig. 9).

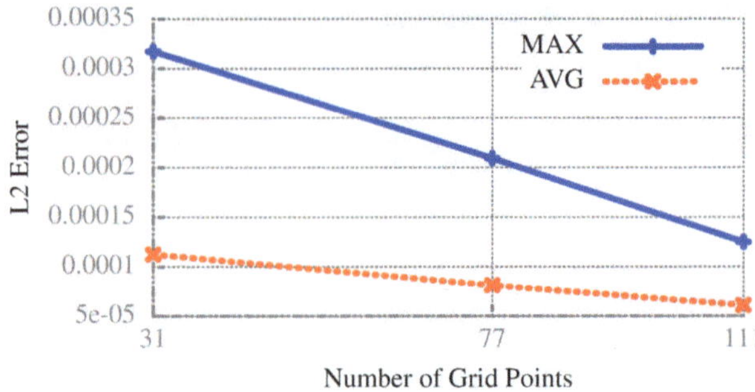

Fig. 9 Accuracy of sparse grid surrogates for the BIM problem for surrogate sizes: 31, 87, and 111. For reference, the maximum velocity values are in the range 5–15 m/s

```
// Initial spvm
for (i = 0; i < m; i++) {
  for (j = 0; j < p; j++) {
    y[j] += x[i] * a[ix[i]][j];
  }
}
```

Listing 1 Serial version of *spvm* (no optimizations)

For our defined simulation scenario, the sparse grid surrogate can fulfill its main purpose of being a cheap exploration indicator by capturing the main features of the simulation. It delivers a good approximation to the original simulation and does so in an almost interactive manner. The current implementation delivers a new interpolated snapshot in $\approx$200 ms.

5 Optimization

5.1 *Sparse Vector Matrix Multiplication (spvm)*

We will now look at the linear combination of snapshots as described in Sect. 4 from the implementation side. Expressed as a linear algebra problem, the task is to compute the inner product of an M-vector x containing m scalar weights ($m \ll M$, i.e. x is sparse) and an M-vector a, whose components contain the snapshots u^{μ_j}, $j \in \{1, \ldots, M\}$ being vectors of length p themselves.

In the following, we will conveniently refer to this task as *Sparse Vector-Matrix Multiplication*, short *spvm*. Listing 1 captures the naïve algorithm for *spvm*. Therein *ix* denotes the compressed vector x of length m.

```
// Auto-tuned spvm
for (j0 = 0; j0 < p; j0 += bs) {
  for (i = 0; i < m; i++) {
    #pragma unroll(uf)
    for (j1 = j0; j1 < MIN(j0 + bs, p); j1++) {
      y[j1] += x[i] * a[ix[i]][j1];
    }
  }
}
```

Listing 2 *spvm* version exposing the tunable parameters *bs* (loop tiling) and *uf* (loop unroll)

spvm is memory-bound, having low computational intensity. The unoptimized algorithm fully loads every contributing snapshot from a, scales it and adds it to y. This involves one multiplication and addition per two memory accesses, i.e. we have 1 flop/memory reference. The optimized version uses tiling of a and y as shown in Listing 2, thus caching the output y which theoretically doubles performance with 2 flop/memory reference. In practice, this performance gain heavily depends on the parameters *bs* and *uf*.

In our interactive setting, we apply the memory-bound *spvm* to large amounts of simulation data, hence our target platform should provide (1) a significant amount of memory (tens of GBs) and (2) high memory bandwidth. Non-Uniform Memory Access (NUMA) systems meet these two requirements as the memory bandwidth scales with the number of NUMA nodes, provided that memory accesses are kept local within each of these nodes. Note that with up to 200 GB/s, GPUs may offer memory bandwidth several times higher than the 50–100 GB/s obtained on a state-of-the-art ×86 CPU-based (NUMA) system; however, the memory limit of ≤6 GB on current GPUs is too low for our purposes. For information and benchmarks for GPUs employed in a similar setting see [3].

5.2 *Empirical spvm Optimizations on NUMA Systems*

spvm is at the core of our whole application and therefore its performance is critical. We speed it up by applying two main optimizations: a NUMA-specific one, and one that is generally applicable. The latter uses tiling on the snapshots in a and the output vector y, such that for sufficiently small tile size the repeated access to y during the reduction operation is done in cache. The nominal flop to memory reference ratio thus increases to 2. The actual performance however is unlikely to double. This is because a so far unremarkable effect is emphasized through tiling and diminishes performance: Jumping from one contributing snapshot (stored in a "row" of linear memory) to the next is unpredictable in a, causing overhead due to indirection. For larger snapshots (and thus larger p), processing the rows of a takes longer and hides this cost. Subdividing one *spvm* into multiple smaller *spvm*s of the same shape, however, introduces more unpredictable jumps, resulting in a trade-off which is difficult to address using theoretical methods.

In order to make use of the increased memory bandwidth of a NUMA system, a multi-threaded implementation of *spvm* must be NUMA-aware. In the k-threaded case we therefore slice the data in a into k equal parts (similar to the tiling above), assigning each part to one thread for further processing. A combination of thread pinning and local memory allocation then ensures that in the following every thread only accesses data placed in the memory of its own respective NUMA node. This behavior is achieved by preparing and executing *spvm* in three steps:

1. Use first touch policy to store slices of a on each NUMA node,
2. Enforce NUMA locality by pinning threads to CPU cores,
3. Execute the multi-threaded, NUMA-aware version of *spvm*.

Note that the one-time overhead caused by the data layout changes in the first step is amortized by multiple invocations of a NUMA-aware *spvm* in the interactive part of the application.

As indicated in Listing 2, we have several tunable parameters in our application: the loop tiling factor *bs*, the loop unroll factor *uf*, and also the number of threads *nt*. The latter is included in the set based on the fact that the optimal value for *nt* does not necessarily equal the number of CPU cores, but can be lower. This is influenced by several factors such as resource contention among the threads sharing the same CPU or trashing the last level cache.

In order to find the optimal combination of these parameters, we employ an *empirical optimization method* (also called *auto-tuning*) [13]. Alternatives are given, e.g., by a theoretical, model-based approach, but in our setting we have conflicting objectives, which often renders analytical solution of such a problem impractical. We want to maximize cache reuse while minimizing the impact of the jumps between rows, however, it is difficult to build an appropriate perfomance model that captures the tradeoff between the two.

We therefore use an orthogonal search method [13], in order to maximize our performance metric, the GFlops rate of *spvm* is given by: GFlops $= (2 \cdot m \cdot p/10^9)/\text{time}$. Orthogonal search always determines the best value for one parameter at a time, setting all the other parameters to fixed values. This approach is also used in the ATLAS library [17], a dense linear algebra library optimized by means of auto-tuning. For instance, the best value for *bs* is searched for while *uf* and *nt* are fixed; afterwards, the range of *uf* is searched while *bs* and *nt* are fixed; and finally, the best *nt* is determined for *bs* and *uf* fixed. Such a traversal assumes a weak dependence among the tunable parameters confirmed in our experiments. Its benefit is a significant reduction of the cardinality of the search space, e.g., $3t$ instead of t^3 for a full search, assuming t possible values for *bs*, *uf*, and *nt*, respectively.

5.3 Performance Results

Our hardware consists of a dual-socket NUMA machine with 6 Intel Xeon X5690 cores per socket clocked at 3.47 GHz. The memory bandwidth as determined by

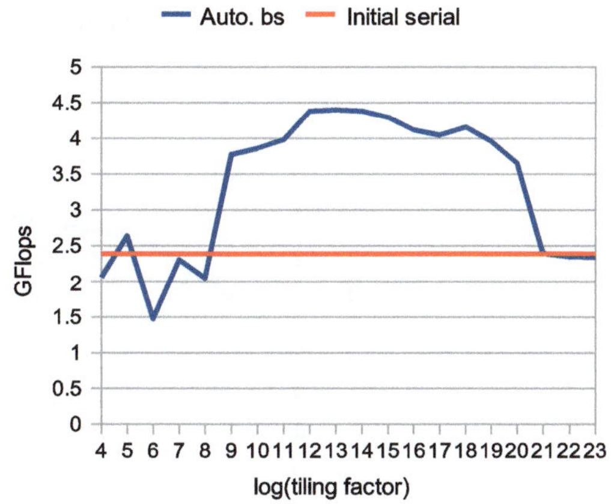

Fig. 10 GFlops as function of *bs*. *uf* and *nt* are fixed

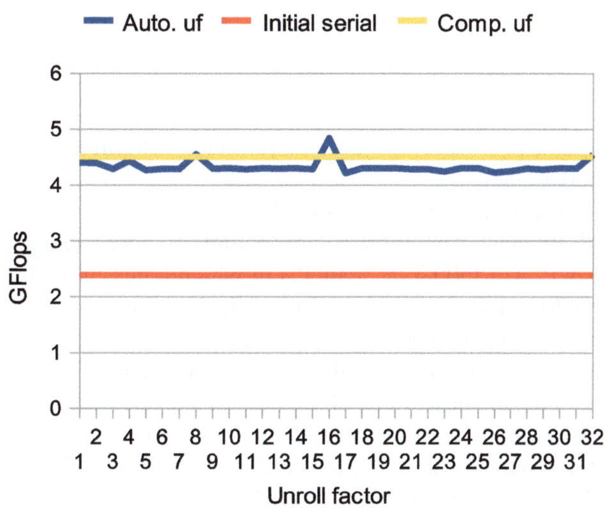

Fig. 11 GFlops as function of *uf*. *bs* and *nt* are fixed

measuring streaming memory access times [9] approaches 24.5 GB/s which limits the GFlops rate of our *spvm* routine to $24.5 * (1{,}024/1{,}000)^3/4(\text{sizeof(float)}) * 2(\text{flop/float}) = 13.2$ GFlops per socket and 26.4 GFlops for 2 sockets. Our input data is in the range of tens of GBs. We use the Intel icc compiler with the flags "-O3 -xHost -openmp". *spvm* operates on single-precision floating point numbers.

The graph in Fig. 10 depicts the performance of *spvm* for different values of *bs*. *uf* and *nt* both have value 1. The reference version (Listing 1) performs almost 2 times slower than the auto-tuned version based on empirical optimizations (Listing 2). The best value for *bs* is at 8,192 (number of floating point values). This value matches to some extent the size of the L1 cache.

In Fig. 11 we can see the dependence of the GFlops rate on the unroll factor *uf*. The best value for *uf* is 16. A notable observation is that the automatic unroll performed by the compiler provides slightly worse performance as shown in the

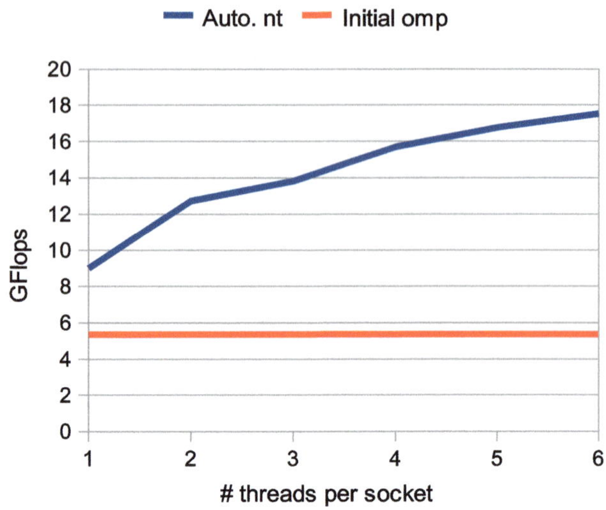

Fig. 12 GFlops as function of *nt*. *bs* and *uf* are fixed

graph. More precisely, by empirically determining the best *uf* we get 7 % better performance.

The graph in Fig. 12 shows the performance obtained using different numbers of threads. The threads are bound to cores using a scatter strategy ("export KMP_AFFINITY = scatter") in which consecutive threads are pinned to different NUMA nodes or sockets. Interesting here is that the best performance is obtained when using the maximum number of threads. This is in contrast with tests performed on other systems: on a dual-socket Nehalem-EP (6 cores per socket, 24 hardware threads) the best *nt* is 8 (4 threads per socket) while on a single-socket Sandy Bridge (4 cores, 8 hardware threads) the best *nt* is 3. Consequently, it is important for performance portability to keep *nt* a tunable parameter. The multithreaded version of optimized *spvm* reaches 70 % of the theoretical peak of 26.4 GFlops. However, on a single-socket Intel Sandy Bridge system, our *spvm* reaches the peak performance of 10.6 GFlops.

6 Visualization

Visualization is an indispensable tool to gain insight into the results of numerical simulations. Especially with large-scale and parameter-dependent simulations as it is the case here, an interactive exploration is the only feasible way to analyze the simulation results: Any change in the simulation parameters, triggered by the user, causes the data that is being visualized to change, and thus the visualization has to be recomputed. Additionally, the very high level of detail in the building infrastructure and the flow around it calls for high-resolution visualization systems such as powerwalls or tiled displays.

Fig. 13 The building and flow data are visualized on a multi-display visualization system. Probes (boxes) serve as seeding locations for streamlines and can be freely positioned

For vector-valued volumetric data, the most common visualization approaches are based on integral curves [2, 8] (called *streamlines* in the case of steady flow), often augmented by direct volume rendering. Our system supports both these options, in combination with displaying the building models themselves to serve as context information. We employ the Equalizer parallel rendering framework [5] to support any visualization system, ranging from a single desktop PC to the largest powerwall and CAVE installations. The visualization system used in our tests is a semi-immersive installation consisting of eight 3D monitors, and can be seen in Fig. 13.

6.1 Streamlines

A streamline $\mathbf{s}(t)$ in a vector field $\mathbf{v}(\mathbf{x})$ starts at a user-defined *seed point* $\mathbf{x_0}$ and is tangent to $\mathbf{v}$ over its parameter interval $[t_0, t_1]$. It is given by the ordinary differential equation

$$\dot{\mathbf{s}}(t) = \mathbf{v}(\mathbf{s}(t)) \text{ and } \mathbf{s}(t_0) = \mathbf{x_0} \text{ for } t \in [t_0, t_1].$$

In practice, streamlines are approximated by using numerical integration methods. To achieve a good balance between integration quality and computation time, we use a fourth-order Runge-Kutta integrator in our work, i.e.

$$\begin{aligned}\mathbf{s}(t + \delta t) &= \mathbf{s}(t) + \delta t \cdot (\tfrac{1}{6}\mathbf{v_0} + \tfrac{2}{6}\mathbf{v_1} + \tfrac{2}{6}\mathbf{v_2} + \tfrac{1}{6}\mathbf{v_3}), \text{ where}\\ \mathbf{v_0} &= \mathbf{v}(\mathbf{s}(t)),\end{aligned}$$

$$\mathbf{v_1} = \mathbf{v}(\mathbf{s}(t) + \tfrac{1}{2}\delta t \cdot \mathbf{v_0}),$$
$$\mathbf{v_2} = \mathbf{v}(\mathbf{s}(t) + \tfrac{1}{2}\delta t \cdot \mathbf{v_1}),$$
$$\mathbf{v_3} = \mathbf{v}(\mathbf{s}(t) + \delta t \cdot \mathbf{v_2}).$$

During runtime, the user specifies the number of streamlines to generate and a seed region (also called *probe*) inside which seed points are placed either regularly or randomly. To gain a quick overview of a data set, the seed region can be set to encompass the whole domain. For a more detailed analysis, a small seed region can be placed near an area of interest. By using the extremely high computational power and memory bandwith of a GPU for streamline computation, we can generate and render thousands of streamlines in real time.

For display, the streamlines are extruded to tubes by a geometry shader. To increase the amount of information in the visualizations, our rendering system also allows the mapping of scalar quantities derived from the flow field to a tube's color and radius. Quantities of interest can either be derived from the velocity field, such as velocity magnitude or vorticity magnitude, or can come from additional scalar fields, e.g. pressure. In the visualizations in this paper, the inverse of the velocity magnitude is mapped to the tube radius, so that the tube gets thicker in slow areas. This allows for an intuitive interpretation of speed even in static images. Some example visualizations showcasing the streamtubes can be seen in Fig. 8.

6.2 Volume Rendering

In order to provide additional context information, our system also provides volume rendering capabilities. Iso-surfaces of scalar quantities (velocity magnitude, pressure, ...), i.e. all points in the volume where the chosen scalar quantity is equal to a user-defined iso-value, can be extracted and displayed. To find an iso-surface, a ray is traced through the volume for each visible pixel on the screen. At regular intervals along the ray, the quantity under consideration is evaluated. A change in the sign of (*quantity* – *isovalue*) indicates an intersection in the previous interval. We then employ a binary search to find the exact intersection location. The gradient at the intersection point is used as the surface normal for illumination. Figure 14 shows a rendering of an iso-surface of velocity magnitude in front of the TUM building.

7 Interaction

An issue that has not been addressed so far is the interaction of the user with our system, e.g. to move the probes from which particles are seeded. With an immersive visualization system, the traditional "keyboard + mouse" interface is not applicable

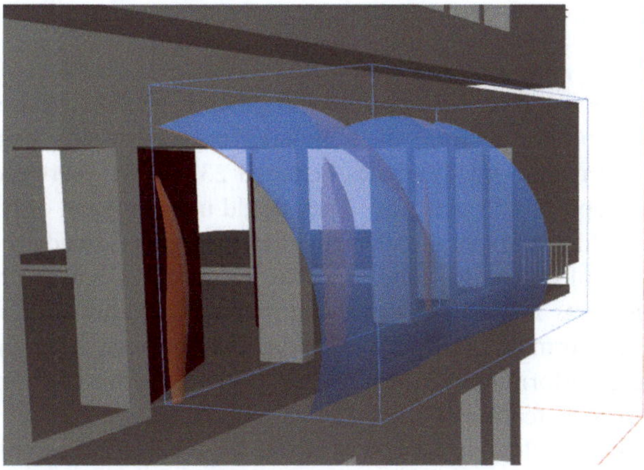

Fig. 14 Iso-surface of flow velocity magnitude for an iso-value of 16 m/s in the region of the parametrized doors. Two-sided shading is employed to distinguish the "outer" side (values < iso-value, *blue*), where the velocity magnitude is below the iso-value, from the "inner" side (values > iso-value, *orange*)

(or at the least, very inconvenient), and as such, alternative interaction mechanisms are required. In the following, we describe the user interactions implemented in our system and analyze their efficiency in a user study. The goal is to allow an easy and intuitive navigation in order to avoid putting any additional cognitive burden on the user. To achieve this, we have implemented an interaction solution based on hand gesture detection using Microsoft's Kinect sensor. As hand gestures are used for interaction in the real world, they offer great promise for an intuitive user interface.

Tang [14] also uses the Kinect camera for hand gesture recognition based on RGB and depth images. He is able to recognize "grasp" and "drop" gestures. Xu et al. [18] use depth information from the Kinect sensor to improve tracking stability and accuracy. Rather than further improving the tracking performance, we investigate the different hand gestures that users perform for different tasks. Moreover, we aim to determine whether or not there is a common mapping among the users between the performed hand gesture and the desired task. In our tests, we use the OpenNI Framework [10] which allows the rapid development and evaluation of virtual reality applications. We integrate the hand gesture detection (wave, push, swipe, circle, and steady) with the skeleton detection to improve the robustness of the hand gesture recognition. For example, if a user performs a circle gesture with the right hand, it is expected and checked that the user also performs a circle gesture with the right shoulder.

7.1 User Study

To investigate the mapping between a hand gesture and the desired task in the virtual environment, we perform the following user study. We explained to the 10 participants (6 males and 4 females between 23 and 31 years old) the desired tasks they were supposed to perform . They could use one hand or both hands and perform gestures like horizontal and vertical swipes, push, steady, circular or any combination thereof under the condition that the gestures have to be unique per task. The participants first had to observe a prerecorded animation of the desired tasks, and then perform the gestures they thought would match the task. While the participants were performing the gestures, the task animation was running again to give the participants a feeling of control.

We determine the following set of tasks to be necessary for the exploration of our simulation results:

1. **Translation of a probe**: Horizontal and vertical movement of a probe (i.e. seeding region) to explore different areas of interest.
2. **Rotation of the domain**: Rotation around the X/Y/Z axes to explore the domain from different perspectives.
3. **Scaling of the probes**: Scaling to control the size of the seeding region.
4. **Introducing and removing probes**: Exploration of multiple areas of interest.

Both the performed hand gestures and the skeleton movements were recorded using the Kinect sensor for post-analysis. For illustration, Fig. 15 below show the 3D hand paths and skeleton movements for a gesture performed by two participants for the task of scaling down.

7.2 Results and Discussion

For the horizontal translation of the probe as an example, we notice that two participants used a circular movement of the right hand around the z-axis, while four participants swiped the left hand in the x-axis for the same task. The other participants used the same gestures but with a different hand (left or right), or with both hands. Even though the majority of the participants used the same gesture, we can not determine a common hand gesture for this task. However, for the task of scaling the probe down or up (see Fig. 15), all participants used the same hand gesture: a hand pinch (bringing the hands closer together or further apart, respectively). This gesture is the most performed gesture in the user study.

The user study shows that the users did not use the same hand gestures for the same tasks, indicating that there is no unique or common mental mapping between the hand gestures and the tasks for most of the tasks. However, the users explained that they used hand pinching for scaling by the fact that they could map this task to the familiar task of zooming on their multi-touch handheld devices. Therefore,

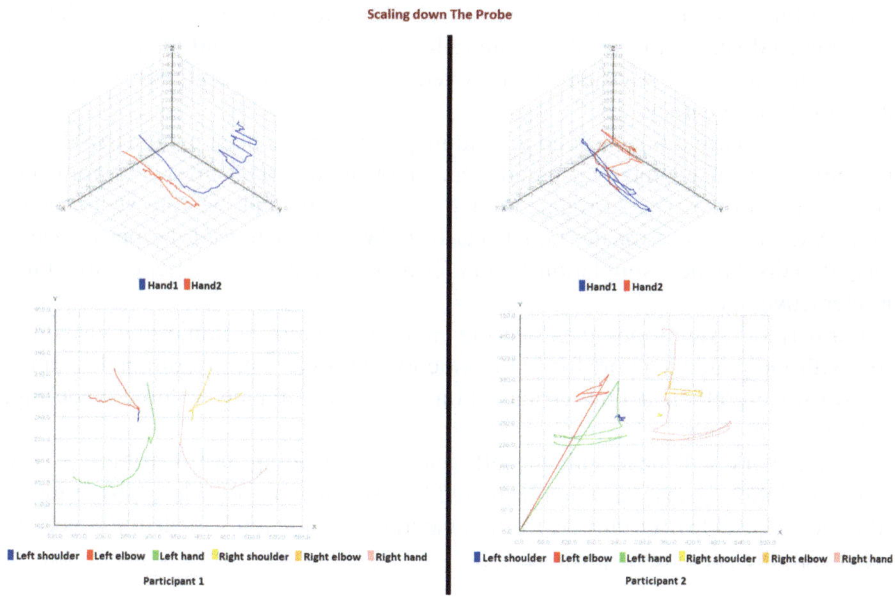

Fig. 15 Scaling down of the probe by two participants

they simply replace finger pinching by hand pinching. We can then conclude that it is more likely that we could reach a unique and common mapping between the hand gestures and the tasks if the users could map the simulation tasks to some familiar tasks they are used to. Exploration of simulation results is unfamiliar to the users, so there is no preconception of the matching hand gesture for a task. We are therefore planning to repeat the user study with a more targeted group of people who are familiar with numerical simulations and their visualization. We also notice that the users mostly performed simple swipe and circular movements using different or both hands to make the mapping gesture-task unique. Therefore, in the next user study, we will choose a set of gestures to perform the different tasks, and ask the users to perform the specific explained gestures. The objective will be then to investigate the robustness of the gesture recognition software for inconsistent and non-expected gestures as well as the users' acceptance of the proposed gestures.

8 Conclusion and Future Work

This work introduced an integrated solution for computational steering of parameterized flow simulations in complex, large-scale geometries. A hierarchical data structure concept was used to efficiently access and generate parametrized simulation geometries from BIM databases. For simulation steering we employed the sparse-grids-based surrogate model as a substitute for the costly simulation.

By investing computational effort in an offline phase, we were able to deliver approximated simulation results in the online steering phase and display them on a multi-display scalable visualization system even for high resolution simulations and with good accuracy.

As test scenario a flow around a building geometry has been parametrized with a combination of geometrical parameters (door angles) and physical parameters (inflow conditions). The surrogate model was examined and the core vector operations were optimized for the target platform. We thus achieved the performance needed to display new simulation results for non-simulated parameter combination at interactive rates.

Furthermore, we proposed as steering environment a semi-immersive visualization with integrated support for 3D interaction with the data. Additional features described were efficient flow visualization via particle tracing, volume rendering, and flexible probing.

Future work will deal with the efficiency of reloading and distributing very large simulation snapshots when triggered by user interaction. Also, designing an intuitive and usable user interface which incorporates domain knowledge is needed to enhance the exploration process.

Acknowledgements This publication is based on work supported by Award No. UK-C0020, made by King Abdullah University of Science and Technology (KAUST). Furthermore, financial support from the German Research Foundation (DFG) within research group 1546 – Rechnergestützte kooperative Trassenplanung in mehrskaligen 3D-Stadt und Bauwerksmodellen (3DTracks) is gratefully acknowledged.

References

1. Bungartz, H.-J., Griebel, M.: Sparse grids. Acta Numer. **13**, 147–269 (2004)
2. Bürger, K., Schneider, J., Kondratieva, P., Krüger, J., Westermann, R.: Interactive visual exploration of unsteady 3D flows. In: Eurographics/IEEE VGTC Symposium on Visualization (EuroVis), Norrköping (2007)
3. Butnaru, D., Buse, G., Pflüger, D.: A parallel and distributed surrogate model implementation for computational steering. In: Proceedings of the 11th International Symposium on Parallel and Distributed Computing (ISPDC), Munich (2012)
4. Butnaru, D., Peherstorfer, B., Pflüger, D., Bungartz, H.-J.: Fast insight into high-dimensional parametrized simulation data. In: 11th International Conference on Machine Learning and Applications (ICMLA), Boca Raton (2012)
5. Eilemann, S., Makhinya, M., Pajarola, R.: Equalizer: a scalable parallel rendering framework. IEEE Trans. Vis. Comput. Graph. **15**, 436–452 (2009)
6. IFC2 × 3 TC1 Release: IFC specification. http://www.buildingsmart-tech.org/specifications/ifc-releases/ifc2x3-tc1-release/summary
7. KAUST: Shaheen. http://www.hpc.kaust.edu.sa
8. Krüger, J., Kipfer, P., Kondratieva, P., Westermann, R.: A particle system for interactive visualization of 3D flows. IEEE Trans. Vis. Comput. Graph. **11**(6), 744–756 (2005)
9. McCalpin, J.D.: Memory bandwidth and machine balance in current high performance computers. In: IEEE Computer Society Technical Committee on Computer Architecture (TCCA) Newsletter, pp. 19–25 (1995)

10. OpenNI consortium: OpenNI SDK 2. http://openni.org
11. The OpenFOAM Foundation: OpenFOAM 2.1.1. http://www.openfoam.org
12. Pflüger, D.: Spatially Adaptive Sparse Grids for High-Dimensional Problems. Verlag Dr. Hut, München (2010)
13. Seymour, K., You, H., Dongarra, J.J.: A comparison of search heuristics for empirical code optimization. In: Proceedings of the IEEE International Conference on Cluster Computing, Third international Workshop on Automatic Performance Tuning (iWAPT 2008), Tsukuba, pp. 421–429. IEEE, Piscataway (2008)
14. Tang, M.: Recognizing hand gestures with Microsoft's Kinect. Technical report, Department of Electrical Engineering, Stanford University (2011)
15. Varduhn, V., Mundani, R.-P., Rank, E.: Real time processing of large data sets from built infrastructure. J. Syst. Cybern. Inform. **9**, 63–67 (2011)
16. Varduhn, V., Mundani, R.-P., Rank, E.: A framework for parallel numerical simulations on multi-scale geometries. In: Proceedings of the 11th International Symposium on Parallel and Distributed Computing (ISPDC), Munich (2012)
17. Whaley, R.C., Petitet, A., Dongarra, J.J.: Automated empirical optimizations of software and the ATLAS project. Parallel Comput. **27**(1–2), 3–35 (2001)
18. Xu, W., Lee, E.: Gesture recognition based on 2D and 3D feature by using Kinect device. In: The 6th International Conference on Information Security and Assurance, Shanghai (2012)

Experiences with a Flexibly Reconfigurable Visualization System on Software Development and Workplace Ergonomics

Marcus Tönnis, Amal Benzina, and Gudrun Klinker

Abstract Physically reconfigurable visualization systems bear the potential to provide more flexibility for content presentation w.r.t. different types of users. Screen setups with alterable layouts allow switching between power-wall and CAVE-like systems, and allow any other setup no matter if display components are totally decoupled or arranged in proprietary setups.

At Technische Universität München we set up such a flexible system built of consumer market 3D displays and state-of-the-art computer hardware providing means for such physical flexibility.

With such a system at operation for a longer time by now, we are investigating the question whether the grade of flexibility has an influence on how software developers adapt to this flexibility. We especially investigate ergonomic aspects of workplace setup of the developers themselves.

Experiences and findings lead towards relationships between display flexibility and developer-related effects, such as the number of concurrent developers and postural deficits developers are willing to endure. This article reports on the flexible visualization system, the determined relationships between its flexibility and postural effects on developers, and discusses attempts to avoid an unbalanced level of flexibility that leads to non-ergonomic states.

Keywords Virtual reality • Human factors • Developer-centered design

---

M. Tönnis (✉) · A. Benzina · G. Klinker
Fachgebiet Augmented Reality, Fakultät für Informatik, Technische Universität München, Boltzmannstraße 3, 85748 Garching b. München, Germany
e-mail: toennis@in.tum.de; benzina@in.tum.de; klinker@in.tum.de

M. Bader et al. (eds.), *Advanced Computing*, Lecture Notes in Computational Science and Engineering 93, DOI 10.1007/978-3-642-38762-3_11,

1 Motivation

Visualization environments gain an increasing importance. The larger the amount of data, the higher might be the need for a suitable visualization. While physical simulations often bear the inherent demand for spatial visualization, other disciplines strive for multi-dimensional data aggregation to gain suitable information visualization properties.

For many types of visualization one-screen desktop systems suffice, for others, large scale and large field-of-view visualization systems are required to provide additional props for visualizing larger portions of data. Spatial data and volumetric simulations define the very end of a field-of-view scale as they integrate virtual worlds which often demand a certain minimum level of immersion.

To set up a visualization system providing the most suitable level of fidelity for the desired area of application requires several trade-offs between building infrastructure (i.e. room size, power, AC, options for darkening), available funding and available facilities for maintenance and operation. One aspect, however, often is neglected. The major focus usually lies on the end-users and their demands – the developers and their circumstances are often disregarded. In contrast to conventional one-screen desktop systems that can be arranged according to ergonomic work guidelines, large scale systems usually require the developer to cope with the physical setup of the system components. At least during the testing and deployment phases, the developer has to work at the visualization system directly. This can cause developers to work in non-ergonomic postures.

We built a visualization system called *FRAVE* to bridge the gap between flexibility of screen placement and the field of view into the virtual world. After almost 2 years in operation, we started looking back to investigate how developers are working with the system. This is the focus of this work: to explore, how our level of flexibility behaves w.r.t. software development in general and with ergonomic aspects in detail, and what we can learn from the findings to provide future users with ergonomically supportive systems. We investigate the question how developers use the system in a physical and postural context, and how the usage affects properties of ergonomic workplace setup. Feedback of users covering their workplace setup is analyzed and discussed w.r.t. risks for potential development of negative postural habits (i.e. musculoskeletal disorders, MSD).

The following sections therefore first introduce the fundamental concept for flexible visualization and then introduce the our visualization system, the FRAVE. In the succeeding part, implications of the physical structure of the visualization system on approaches of software developers and their organization of their workplaces are investigated. The main focus lies on the analysis of a questionnaire collecting data about workplace setups. Observations, among other things, show that the flexibility of the setup pipes through to the developers in general. The analysis of the collected data shows that special care must be taken to ensure ergonomically health-preserving system use.

2 Flexible Visualization Environments

Virtual reality for its own purpose and the demand for visualization of simulation data generate an increasing importance for various and varying types of visualization environments. While many virtual environments require fully immersive visualization systems with necessarily high levels of fidelity for sufficient presence, others require to satisfy fewer requirements. Simulation systems of spatial and physical properties, for instance, often require lower levels of immersion. Researchers in such fields yet require a wide field of view and high resolution to investigate spatially distributed properties of the simulation. Besides knowing about the spatial context and having met the demand for interfaces to travel to the specific location, visualizing the according portion of the simulated space might be sufficient to such users. The focus is not to get into the virtual world mentally, it rather is to understand the occurrences and effects of the simulation.

As such an understanding might be strongly based on spatial relationships, it thus does appear important to not let the researcher deal with spatial transformations in a larger extent than necessary. When traveling through a virtual world or when examining a point of interest from different perspectives, the user in fact has to maintain the viewpoint location and orientation based on virtual cues rather than on physical cues. The user simultaneously has to maintain his physical motion and the mapping of the virtual world w.r.t. the physical world. Physical cues indeed might be supported in the walkable restricted space of a CAVE, but orientation changes when the virtual world is rotated with a user interface. Bowman et al. [5] states that each 60° rotation requires 1 s to mentally execute. To the end, two different worlds require spatial maintenance the same point in time. Mounting the virtual world to the real world – at least in those situations where examination tasks of simulated data are conducted – bears the potential to let users reduce, or even neglect, such transformational issues.

We were searching for an option to let users of visualization systems have a more flexible approach to deal with their position and orientation in a virtual world (as said, at least for the part in which they are examining a specific occurrence of a simulation). Creating a partially rigid connection between the physical and the virtual world stated a suitable support. Such a connection actually already exists. The head-tracking dependent frustum computation w.r.t. the displays lets the screens become spatially fixed windows into the virtual world. We yet emphasize this fact and extend the way a user can make use of it.

This rigid connection and its usage shall be explained on a generic example. A CAVE might be used to visualize a virtual environment generated through a simulation. The simulation computes and visualizes its data, bearing an effect requiring further investigation. A guidance system might show a path to this location and the user employs some interface to travel there and eventually reaches the point of interest. Here, the virtual world moves w.r.t. the physical world. The world stays in place as soon as the travel interface is released. A general CAVE yet provides another interface to maneuver spatially: the beforehand mentioned head-tracking facilities – even if this facility restricts to the limited area of the CAVE. Both, the user interface and the head-tacking facilities affect the position and orientation of the user's viewpoint in the virtual world, but only the head-tracking system maps physical motion and thus enables truly physical cues. The tracking system, in case of head-tracking, does establish the direct connection between physical and virtual world. Assuming that the CAVE does provide four side walls and that the area of interest is smaller than the accessible space inside the CAVE, the user will be able to walk around the interesting region and thus will be able to investigate the simulation result from all directions. All physical motions are then directly mapped to the virtual world.

This example could inherently define the demand for visualization systems to provide a walkable space larger than what a common CAVE does provide today. But why should such a system be a CAVE at all? Would a set of flexible display components that can be arranged according to user preference be sufficient as long as the screens provide the view in whatever direction is required?

As described in more detail in a technical note [13] we combine panel displays with portability features to yield such a flexible visualization system. Although they do have bezels reducing immersion, they are easier to handle than projector-wall systems. Especially the physical dimensions of panel displays enhance mobility, and their depth is significantly shorter than any arrangement of a projector and projection wall setup.

The next section investigates different approaches for visualization systems. This is followed by an illustration of our concept. Subsequent to the general concept is a detailed discussion of hardware specific trade-offs we had to deal with when installing our FRAVE system.

2.1 Related Work for Visualization Systems

While most of the traditional larger-scale visualization systems employ projectors and projection walls, using panel displays for visualization systems is not a new strategy at all. The NexCAVE [6] showed that panel displays can be used to build CAVEs. The setup uses tight and rigidly mounted displays. All screen-normals are oriented towards a common center point, the spot from which the generated picture is perceived best with minimal bezel obstruction. Using projection systems, but providing some flexibility, the RAVE from Hindus [9] allows the side walls to be turnable around the inner joints of a three-sided setup. The tiltable wall by Doulis et al. [7] enables the user to adjust the screen setup according to personal preferences. This can be compared to the approach of Bimber et al. [3] who foresaw a seamless integration of VR systems into habitual office workspaces.

Rekimoto [12] uses several independent displays, defining each display as another window into the virtual world. Rather than changing orientation and position with interface devices, users locate the display screens so that they are facing in the required directions. The virtual world itself can be moved for larger travel tasks, but remains fixed w.r.t. the physical world when smaller, physical movements are performed. This approach essentially establishes the paradigm of Augmented Reality, maintaining the requirements defined by Azuma [2], registration in 3D and interactivity, and all in real-time.

2.2 FRAVE Concept

We loosen the tight, rigid connection between displays such as in the NexCAVE, making the display setup reconfigurable while keeping the opportunity of a CAVE-like setup available. More generally and following Azuma [2], we integrate the virtual world into our physical world when moving physically. This allows users to perform near field maneuvering without having to translate the virtual world iteratively. Users thus do not have to memorize positions and alignments of spatial relationships in the virtual world to keep track of their path and position – they rather can rely on physical cues. All spatial dependencies can be tracked according to physical steps, the motoric memory and physical awareness. Users yield different views by moving displays and looking through these onto the virtual scenery. The displays themselves become the input and steering devices for viewpoint control. Such a visualization system requires tracking capabilities to track the displays and the head position to compute a correctly aligned field of view for each display. The displays become portable windows into the virtual world.

Our proposed system, the *FRAVE*, a *Flexibly Reconfigurable CAVE* was intended to be such a flexible system. The system is based on the vision of a physical environment where users set up their visualization displays according to personal preferences and demands. Multiple displays can be combined to a user-enclosing 3D environment similar to a CAVE. Display elements can be arranged freely, only depending on the available space. Examinations of sceneries in any direction are facilitated, even “inside” facing displays are possible. The user does not have to reposition the virtual within the physical world (when the virtual area of interest is smaller than the available physical space).

With the focus on visualization of simulation data in combination with terrain data, we developed a set of display components that can serve these demands. Figure 1 shows two possible configurations. To some extent these components correspond to the RAVE approach of Hindus [9] where the two side walls of a three-sided panel architecture can be turned around their common joints with the center wall. The complete FRAVE system consists of ten displays in different partial arrangements and with different capabilities. A ground floor facilitates two displays and is covered with glass to be capable of carrying two users. Three wall segments are equipped with two displays each. The walls have rolling bases and can be moved around freely, for instance, into an open position similar to a power-wall (Fig. 1a) or independently to a custom setup (Fig. 1b).

Two other displays provide the highest flexibility. Each is mounted on its own rolling base and is adjustable in height and tilt. Such as with the tiltable wall by Doulis et al. [7], but with more degrees of freedom, these elements can be adjusted

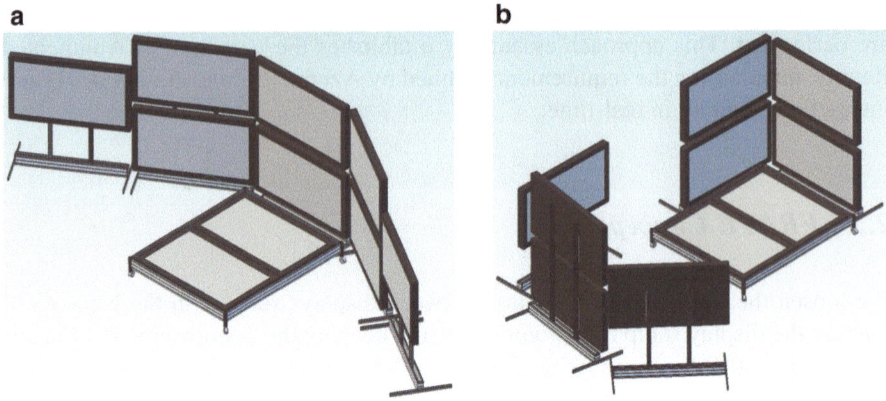

Fig. 1 The FRAVE concept shown in sketches. (**a**) A power-wall like setup. (**b**) A setup of independently placed components, i.e., for two teams

Fig. 2 Adjustable screen in tilted and lowered position

to personal preferences of the users. Figure 2 illustrates a possible configuration. An upright position enables users to stand while perhaps discussing simulation effects. A tilted setup enables users to look downwards. Lowered positioning allows users sitting down for longer sessions.

2.3 Balancing Hardware-Specific Trade-Offs

The original FRAVE concept foresaw comparably large panel displays. To draw a final decision concerning display size and the overall size of the FRAVE some more influencing parameters were necessary to consider. While a more detailed description of all factors on the FRAVE is given elsewhere [13], we here summarize the factors that later require to be discussed w.r.t. effects on software development and ergonomic effects.

Mainly for visualization components a choice in the display size was required. Our aim was to provide a large field of view with as few visual obstruction as agreeable. Generally, smaller displays have smaller bezels. However, the absolute number of bezels of a four panel display array substituting an equally large single display quadrupels. Another influencing factor for the visualization system in general is that an average height user (180 cm) should be able to see some portion of the image above the straight forward horizontal line of sight. As also a ground floor with displays was to be installed, the lowest line of the lowest display should reside directly adjacent to the floor display. These requirements yielded the following differentiation. If the displays are larger than 60 in., two displays, one above the other, are sufficient to cover the necessary height. Between 40 and 60 in., three display are required and below 40 in., four displays are necessary to reach the required height in display space.

Usually smaller and mid size displays are cheaper than the largest available ones. Working with smaller displays yields a higher resolution. However, the calculation of the number of displays has to factor in the cost of computers and graphics cards. As most displays used active shutter technology at the time of construction, the graphics cards driving the system were required to provide functionality for hardware-based synchronization (i.e. G-Sync). Our choice therefore went to displays of at least 60 in. in the diagonal, keeping costs for computers and graphics hardware comparably low. To the end, the choice went to the Panasonic VT20 series with 65 in. diagonal as their displays yielded best quality estimates concerning picture quality, energy consumption, and IR shutter signal transmission angles. A pixel has a diameter of 0.75 mm yielding an angular resolution comparable to current upper end visualization systems. Each wall thus employs two displays.

Each building block (floor, 3 walls, two adjustable single displays) is driven by a separate computer equipped with an NVidia QuadroPlex 7000. Both, the computer and the graphics card are mounted to the backside of the respective frame. Each computer for convenience is equipped with mouse and keyboard. The respective USB cables are equipped with extension cords.

2.4 *Dimensions of the FRAVE*

The final construction had a ground floor at a height of 30 cm and an overall height of 220 cm. Each of the three walls is 160 cm wide. The ground floor, due to the encapsulating frame, is 200 cm wide and has a length (depth) of 230 cm. The glass plates of the ground floor are 39 mm thick and weigh approx. 150 kg each. The lower wall display screens start at a lower height of 40 cm and have the upper end at 120 cm. The upper wall display screens start at a lower height of 135 cm and have the upper end at 215 cm. Each wall is rollable and has an approximate weight

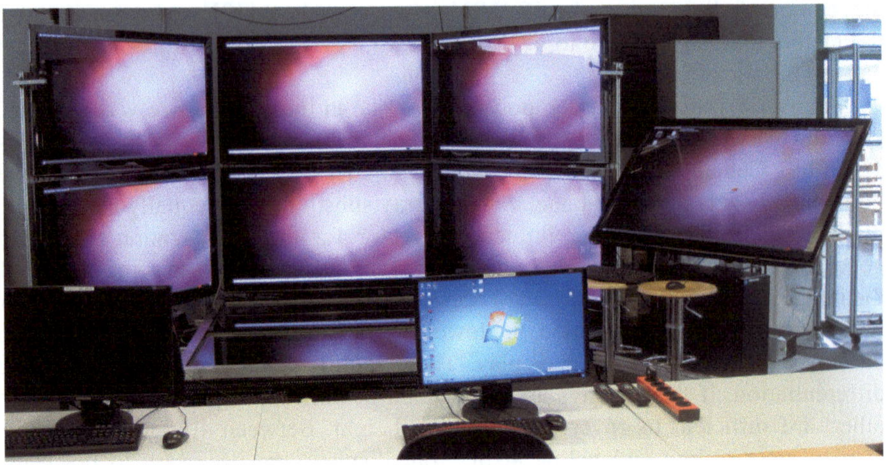

Fig. 3 General FRAVE setup: three walls in bended power-wall setup with adjacent tiltable display and remote desktop computers in front of FRAVE

of 200 kg including the two displays, the computer, the external graphics card and the mounting frame. The most often used configuration has the two sidewalls tilted inwards by approximately 30°, see Fig. 1a.

The two adjustable screens are manually tiltable to any angle between vertical and horizontal and are adjustable in height between 145 and 190 cm (upper edge of screen; lower edge from 65 to 110 cm) with an electronic actuator. The also rollable one-screen systems each have an approximate weight of 100 kg including the display, the computer, the external graphics card and the mounting frame.

We installed another three computers for remote desktop access on a bench of three tables in front of the FRAVE. Figure 3 shows the common setup with two of the three remote desktop computers visible in front. These workplaces were preinstalled according to ergonomic guidelines. The intention behind the setup in front of the FRAVE was to enable developers to gain a straight line of sight on the FRAVE when directly executing applications on the large displays (when the system is set up in the general 30° configuration or totally opened). The tracking system of the FRAVE covers almost the full extent of the desks. Developers thus do not have to stand up for testing.

Ubuntu 10.4 (later, 10.10) and Windows 7 were installed on all computers of the FRAVE (both, visualization nodes and remote desktop computers).

3 Implications on Software Development

With this setup in operation, we started observing how the development infrastructure was used and how developers worked in that environment. This section covers observations concerning the flexibility of the visualization system and how it bears potential to support concurrent development with multiple users in the first part. This is followed by investigating the main question of this work, how the flexibility to dynamically position visualization components and the comparably large scale of the screens affect ergonomic habits of people working with the system.

3.1 Flexible Support for Multiple Developers

While a central intention of the FRAVE was providing flexibility to the end user, we can extend this aim to application developers. Main elements of applications are surely developed on standalone computers, but when it comes to deployment and especially to testing systems and user interfaces, the target platform needs to be entered.

Let us illustrate a development process by taking a look at the highest-fidelity form of visualization systems, a CAVE. Most such systems are equipped with one master control computer standing at a desk outside the CAVE. The developer sits down in front of a desk with a monitor, and may be able to look into the CAVE, perhaps with an inclined glance. All development work has to be conducted from this one computer because the computers of the CAVE are usually not accessible directly. Further computers of course can be set up – at least as long as there is space available from where the developer can look into the CAVE. However, to test the system, one eventually needs to enter the CAVE. Only there, the user can see if the system runs correctly on all nodes and screens and only there one can test if any user interface requiring tracking operates as planned. The operator's desk could be placed in the tracking volume indeed but that would possibly narrow the entrance to the CAVE. To the end, only one developer is able to at least get a good view from the outside as only the opposing wall is visible to a larger extent.

Power-wall setups as a counterpart to a CAVE define a totally different development environment. Multiple users can operate in parallel by splitting the screens in agreement. Yet, to test different branches of the same software in parallel, the system needs to be extended to be able to concurrently deal with different spatial transformations for the used input devices.

With a repositionable system such as the FRAVE, developers are able to work on the same software system at the same time, even if views into different directions are required. They basically place the visualization components w.r.t. their desk (or input devices) so that they can perceive the desired view immediately after startup. Different, but concurrent views are possible this way. This is further supported by the visualization nodes being equipped with their own set of keyboard and mouse, which usually can not be achieved with rigidly coupled visualization components, no matter if it is a CAVE or a power-wall. The only constraints are the number of screens available and the fact that display panels might conceal the field of view of the tracking system. The standard setup of the FRAVE, however, provides a tracking volume ranging up to the default setup of the remote desktop computers.

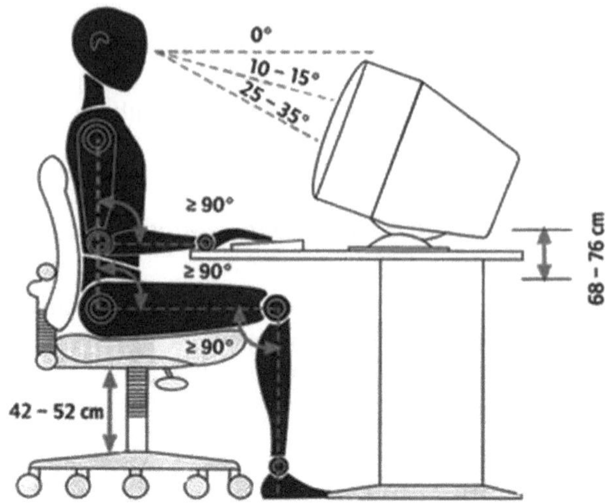

Fig. 4 Workplace ergonomic guidelines (Courtesy of IBB GmbH, Germany)

3.2 Ergonomic Aspects of Health

Flexible visualization systems provide a higher grade of flexibility for developers to establish their working setup. That raises the questions whether the grade of system flexibility relates to the grade of ergonomically healthy software development.

To speak of ergonomically healthy software development requires to briefly discuss the relationship of ergonomic factors in the field of software development or, more general, in the field of computer supported work. Germany and many other countries define sets of guidelines w.r.t. ergonomically correct setup of computer desktop workplaces. Such guidelines are mainly based on the ISO Standard 9241-5 [10]. Further on, the ISO Standard 11064-4 [11] defines ergonomic design of control centers. Figure 4 summarizes the common core guidelines in a sketch. Of major interest in the following discussion are the requirements to have a horizontal line of sight towards the upper edge of the screen and the mostly rectangular (knees, pelvis) or straight (spine) body posture. These factors ensure a health-maintaining working environment and reduce the risk for the development of MSDs.

The design of the FRAVE for the end-user generally ensures these guidelines at least for the upper displays in a standing posture. Our aim is to investigate how the situation evolves for the developers rather than for the end-users. As the incidence of computer-related MSD has been estimated to be around 20 % [8], specialized workplace setups such as in the case of software development in a larger scale visualization environment require investigation of possible influencing factors.

We therefore designed the following questionnaire, also including the collection of common demographic data (age, gender, position at university).

- What is your usual average time per working session at the FRAVE remotely from your work place?
- How many such working sessions are you doing per week in average?
- If you are working from your personal desk, did you adjust the setup (chair height, desk height, screen height, distances, ...) according to computer workplace ergonomic guidelines?
- —
- What is your usual average time per working session at the FRAVE locally in the ITüpferl?
- How many such working sessions are you doing per week in average?
- How many people were present at the FRAVE in average during such a session?
- On which computer do you work? Please note: E.g., logging into frave0 from fraveui0 would yield frave0.
- What is your preferred seating (i.e. chair, wooden stool, bar chair, floor, standing, on the glass floor of the FRAVE, ...)?
- Where is your seating location (i.e. in front of the FRAVE in the middle, left corner of FRAVE floor, ...)?
- On which screen (display device) do you actually conduct your work (i.e. upper screen of frave0, ...)?
- Where do you place mouse and keyboard (i.e. both on desk in front of me, mouse on wooden stool, keyboard on lap, ...)?
- Can you tell, why you did choose this setup (seating, screen, keyboard, ...) (i.e. other computers blocked, did not know about remote desktop login option, ...)?
- Did you feel any kind of back pain or neck stress or other symptoms of non-ergonomic working poses after such sessions? If fine with you, please indicate your problems, otherwise just state yes or no.
- —
- If you have changed your working and seating setup, please indicate why (i.e. other computers blocked, did learn about remote desktop option, changed work focus i.e. from developing to testing, ...) and how your new setup looks like (Note: To answer how, please answer the questions of the marked block again)? Please continue for each setup used for a longer time.

We handed this questionnaire to all users of the FRAVE and received answers from eight users, all except one developing software for the FRAVE. Six users were developers of 3D user interfaces, a domain area which usually requires frequent physical access to the system. To gain further insight into workplace setups and how user seat themselves, we also employed our personal observations from another 11 individuals, all developing either user interfaces, rendering components for the FRAVE or administrating the system. Understandably, no feedback about the second groups' physical exertion could be collected. No statistical analysis of the collected data has been calculated because the feedback contained a too wide variety of different setups to yield meaningful data. We moreover stand in for the fact that already a sole occurrence of non-ergonomic workplace setup requires to be addressed as it bears the potential to evolve into a habit and thus can lead to postural

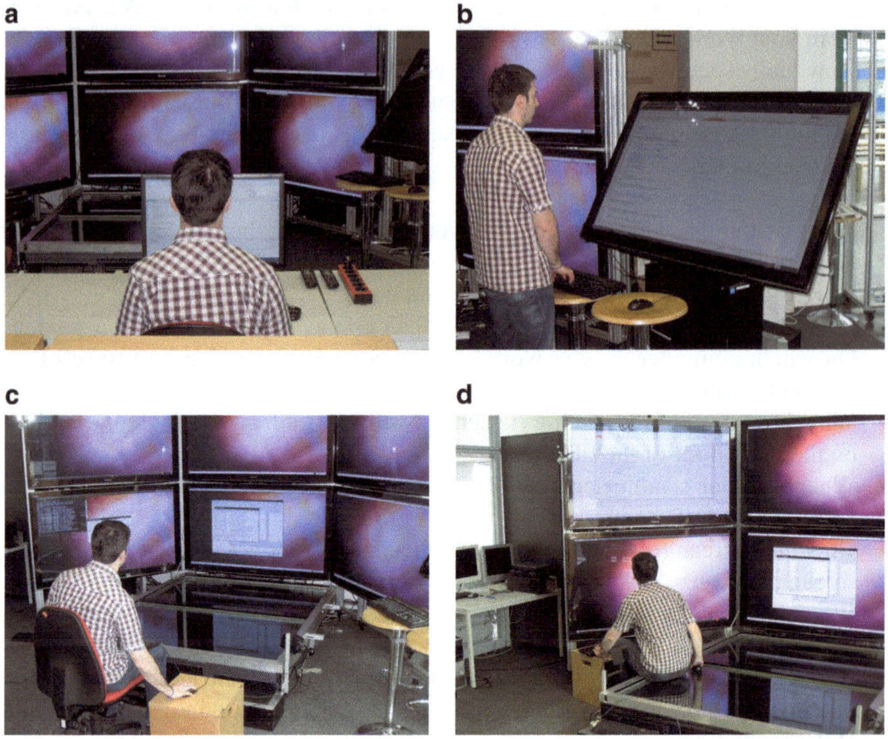

Fig. 5 Four types of working setups at the FRAVE. (**a**) Sitting on a work-desk. (**b**) Standing in front of an adjustable screen. (**c**) Sitting centered in front of the FRAVE. (**d**) Sitting on the ground floor

deficits and MSDs for this one user over time. Any questionable workplace setup is therefore illustrated and discussed to avoid such health-critical usage.

3.2.1 Types of Workplace Setup

An iterative categorization lead to four general types of workplace setups among the developers. The categorization was mainly driven through the physical location where the developers executed their work. Variations of interaction device placement were fed into subcategories. Figure 5 shows the four types: sitting on a work-desk, standing in front of an adjustable screen, sitting centered in front of the FRAVE, and sitting on the ground floor. We discuss these four types in ascending order of negative influence on body posture.

Sitting on a Work-Desk

Users are sitting at a desk with conventional monitor, keyboard and mouse. Figure 5a shows that the FRAVE screens are mainly visible above and besides the monitor on the desk. The users mostly have the display of a visualization node exported to the desktop computer. Sometimes they are running the development environment on one of the FRAVE screens directly. In fewer cases users were sitting at their personal desks elsewhere in their offices.

Developers used this setup mainly for longer coding sessions. Testing and parameter tweaking activities such as with spatial user interfaces were also conducted to a larger degree with this setup. It seems that providing the possibility to use tracked hardware from a seated position proved beneficial for maintaining a harm-free posture.

This setup suits best to an ergonomic workplace setup. The operational setup more or less, but commonly to a larger degree, followed setup guidelines for workplace environments.

Standing in Front of an Adjustable Screen

Figure 5b shows another setup being used especially when working on the tiltable and height-adjustable screens. The developer is standing in front of the screen, using input devices placed on bar chairs. Bar chairs were available as there was a kiosk environment nearby.

Here, the risk for developing a bad postural habit is still comparably low as the users constantly change position because they are standing. The neck is straight when looking at the screen and the input devices can be reached almost comfortably, even if too low.

To be noticed is the usage of bar chairs and the fact that the screens are tiltable, height-adjustable and can be moved around. Almost never has any capability to adjust the screens and the base location been used. Only two exceptions were observed. First, when the screen has completely tilted to a table or when one of the two screens was taken to another office for some time. The new users took the advantage to adjust the configuration to their preference.

It might come from the need to execute physical work that the screen was tilted so seldom, even if adjusting the screen in height just required pressing a cable-bound remote control for lifting. If the lifting and tilting features were used, the adjustable screen could be brought down and tilted in a way that a user could operate in an ergonomically better posture. Then, either an easily transportable elevated desk could be brought to the system or the system could be moved to a location in front of an exiting desk equipped with a remote computer. To enable higher usage rates of these features, we assume that the tilting facility needs electronic support and that the controls need to be prominently placed to reach higher noticing rates by the users.

Concerning the rolling capabilities we noticed a drawback of the construction. The rubber rolls have a comparably low diameter and width. With the one-screen system having a weight of almost 100 kg moving around is indeed not troublesome but requires some effort. This finding might appear of lower influence but the ability to move something easily (through larger and harder rubber rolls) would increase its usage in consequence.

A further handicap for relocating visualization components is cabling. The number of cables were reduced to the necessary minimum. Still a power and a network cord limited position flexibility by blocking the way and partly by being stuck elsewhere, thus not being able to be pulled as required. Ceiling mounted cable conduits at swivel arms surely would eliminate this issue. Yet, installing such facilities for multiple independent components requires additional efforts that are usually beyond the scope of computer scientists.

Another approach could be to mount a drawable desk panel for keyboards and mouse to the screen. Then, however, the user would be sitting too near w.r.t. the size of the screen.

Sitting Centered in Front of the FRAVE

The third often used setup was mostly used by developers who had to repeatedly access the FRAVE directly. Such users took a chair, put the keyboard on their lap and the mouse either on a wooden stool besides or just on the ground floor in front. Figure 5c shows such a typical setup.

Several reasons exist letting users take this setup. The joy working on an unusually large system with screens one would like to have at home was a strong factor letting people sit like this enjoying what is currently their system. Other people took that pose because they just did not know about possibilities to remotely log in to other computers. Sitting in front of the FRAVE, all keyboards and mice of all FRAVE computers could be accessed quickly. Some few people also did not know that the tracking volume also covers the desk space of the remote desktop computers.

Especially as comparable setups were used for quite long durations, we could see an elevated risk for postural deficits. Having the mouse in front on the floor of the FRAVE requires users to lean forward. The users had to change position at least sometimes. Having the mouse on a low platform besides the body bears the potential for long unchanged postures. As observed, users either bend sidewards into a pose where the spine is unevenly balanced laterally or slipped down on the chair into a crooked posture to reach the mouse. Concerning neck bending, we observed that users mostly worked on a lower FRAVE screen having a straight neck and a horizontal line of sight onto the upper edge of the screen, both as recommended in ergonomic guidelines. However, some developers used to work on the upper screens sometimes, then bending their neck comparably wide backwards.

Sitting on the Ground Floor

Figure 5d shows the most concerning posture found among the analysis of the usage. Surprisingly, this setup was used quite frequently. The user is sitting on the FRAVE floor directly in front of the wall screens. The keyboard is either on the user's lap or on a wooden stool. The mouse is either on a second wooden stool or besides the user on the floor of the FRAVE. Usually the lower screen was used, but in rare occasions we observed usage of the upper screen.

The computer of the left FRAVE wall by convenience is the head node for applications running on the FRAVE. That might be the main reason why many users took a seat on the left side, but other users worked on the other walls in similar setups. All types of work had been executed in this setup. Besides system administration, long coding sessions were held, and integration tests, debugging and parametrization of user interfaces was performed.

At the very beginning of the physical deployment of the visualization infrastructure, no extension cords for the USB cables of mouse and keyboard were installed. For a couple of days administrators had to use such a posture to install the operating systems and to configure the system. Usage of this type of seating yet continued even after the extension cords had been installed and after the remote desktop computers were set up. Our main explanation is that new people coming into the room saw how users were seated and continued with this habit.

Such a posture comes with several concerns. The spine is crooked in this crouched stance. The spine is twisted to let the user look on the screen or to reach the mouse besides. The neck is tilted backwards as seating height is too low w.r.t. the screen. The screen is too near, enforcing the user to focus to short distances and to turn the head extensively to gain sight on the left and right areas of the screen. All these factors bear an additional risk for neck stress as it is already bent back. It thus is not surprising that neck strain and back pain were reported.

3.2.2 Discussion

The findings from the questionnaire and the observations imply some major reasons for the use of non-ergonomic workplace setups. First, lacking knowledge of developers about software tools for remote login. Second, potentially missing awareness about the need for ergonomically correct workplace setups. Dependent on the awareness is the physical presence of suitable, flexible setups of the equipment and for desks. These aspects and the interventions proposed earlier, correlate with the most frequent approaches from the survey by Boocock et al. [4]. We discuss all aspects and options to emphasize the demand for maintaining ergonomic postures for future health.

Software Tools

Remote login tools exist for any operating system. Users can log on to other computers, can export the display and even graphics contexts. The knowledge about the availability of such features and their usage needs to be shared among users, especially to new users. This a demand for system operators and project leaders.

Starting a new development project usually comes with lots of new matters for the new developer. Any additional information thus might be forgotten quickly, especially if it is not of direct concern to the actual development task. Two approaches should bear potential to ensure sustainable use of remote login tools.

First, the introduction phase to the visualization environment should be used to set up the working environment already in a way that ensures the use of remote desktop tools. Small workarounds such as "Uh – let me just do this one thing directly there" should be avoided. Getting into a non-ergonomic pose once may often lead to the effect that the user continues with it.

Second, emphasis should also be invested on options preventing the demand to log on to a system physically. Standard settings with logon screens require a user to be physically logged in to a system to run graphical applications on the corresponding screens. If the system can not be configured to not require a login at all, maybe due to demands of access restriction, settings should be enabled to allow running graphical applications on top of the login screen. This of course should be explained to developers, too.

Awareness about Potential Health Issues

At university, developers are usually younger and might not yet think of potentially unhealthy seating and workplace setups. Non-ergonomic workplaces however can already lead to habits of unhealthy postures. This fact should be taken as a motivation to argue why it might in many cases be valuable to use remote login features instead of getting into arguably unhealthy poses. Giving such an additional rationale bears the potential to strengthen the willingness of users to use facilities that might not appear straight forward. Especially, if the facilities to log in remotely are provided, such an argument bears the potential to weigh towards health-maintaining postures.

Flexibility in all Contexts

Any intended flexibility needs to address any context, not just visualization.

This starts with the flexibility of the visualization system. Providing, for instance, the option to roll displays around must be enabled in a way that it is easily executable by users. Two facts delimit this with the FRAVE in general, the rolls being too small and the cabling. A third fact w.r.t. the tilting option of the two extra screens is the required manual work. The efforts to address such issues should not be underestimated. While it is comparably easy to construct mobile mounting frames for displays, the complexity to provide an easy to handle flexible setup requires careful handling of different trade-offs between all components.

A second context of flexibility is the actual executive working environment of the user. While much work can be ported to remote desktop systems, other work requires the developer to leave the remote desk to enter the visualization system.

Then, facilities must be provided to enable the user to easily set up a desktop environment to work from the currently required position. Providing suitable desk setups with the same flexibility as the visualization components is a necessary demand. This correlates to the findings of Akanbi and Nkechi [1], the workplace setup needs to face the demands for accurate postures in both, desk and chair adaptivity.

4 Summary

Flexibly configurable visualization systems bear the potential to let researchers better maintain spatial relationships as they provide physical means to cognitively relate a computer generated virtual world to the physical world. The FRAVE represents such a system, providing several dynamically and independently positionable display components. To test whether such a system does enhance such factors, dedicated software systems need to be developed. The development process lead to the question whether the flexibility of the system enhances flexibility for the developer, too. Dependent on the flexibility, the central question of this work was to investigate if this flexibility leads to developers still maintaining ergonomically correct workplace setups.

Our investigation showed that developers not necessarily tend to explicitly care about their postural habits. The flexibility of the system has only been used partly. Our investigation showed four differentiable types of workplace setup in the FRAVE environment bearing different levels of non-ergonomically postural habits. Investigating the underlying reasons for such behavior, we addressed three main factors requiring care to be taken. It has to be a responsibility of the system operators to (repeatedly) inform the users, especially new users, about both, software and (mechanical) hardware related capabilities. Besides getting informed by others, developers need to develop a level of awareness about the risk of postural deficits.

The last conclusion is that hardware setups for flexible systems require a comparably equal amount of engineering work as is usually put in the software infrastructure. The attempt providing a flexible visualization system does not only require physical decoupling of screens. Besides providing flexibility for screen setups, more flexible setups for work-desks, such as adjustable standing desks, must be provided. Also physical constructions of the visualization system itself require to meet high demands on flexibility and handy usage. Facilities such as electrical cable conduits for multiple mobile devices are necessary to install, but are usually beyond the scope of researchers in the field of computer science. It does require profound knowledge in mechanical engineering to get to a system that is not just to be called flexible but that really is flexible and still let's users execute their work in health-maintaining workplace settings.

References

1. Akanbi, O.G., Nkechi, I.A.: Tailor's chair and Musculoskeletal disorder in Nigeria. In: Proceedings of the International Ergonomics Association (IEA), Beijing. CD-ROM Proceedings (2009)
2. Azuma, R.: A survey of augmented reality. Presence-Teleoperators Virtual Environ. **6**(4), 355–385 (1997)
3. Bimber, O., Encarnação, M., Stork, A.: Seamless integration of virtual reality in habitual workplaces. Zeitschrift für Arbeitswissenschaft **55**(2), 103–112 (2001)
4. Boocock, M.G., McNair, P.J., Larmer, P.J., Armstrong, B., Collier, J., Simmonds, M., Garrett, N.: Interventions for the prevention and management of neck/upper extremity musculoskeletal conditions: a systematic review. Occup. Environ. Med. **64**, 291–303 (2006)
5. Bowman, D., Kruijff, E., LaViola, J.J., Poupyrev, I.: 3D User Interfaces: Theory and Practice. Addison Wesley, Harlow (2004)
6. DeFanti, T., Acevedo, D., Ainsworth, R., Brown, M., Cutchin, S., Dawe, G., Doerr, K., Johnson, A., Knox, C., Kooima, R., et al.: The future of the CAVE. Cent. Eur. J. Eng. **1**(1), 16–37 (2011)
7. Doulis, M., Wyss, H.P., Rietmann, M., Agotai, D.: The TiltingWall – a highly flexible virtual reality test environment. White paper http://www.i4ds.ch/uploads/media/TiltingWall_02.pdf (2008). Accessed 6 Sep 2012
8. Gerr, F., Marcus, M., Ensor, C., Kleinbaum, D., Cohen, S., Edwards, A., Gentry, E., Ortiz, D., Monteilh, C.: A prospective study of computer users: I. Study design and incidence of musculoskeletal symptoms and disorders. Am. J. Ind. Med. **41**(4), 221–235 (2002)
9. Hindus, L.: Immersive 3D engineering environments are all the RAVE (Reconfigurable Advanced Visualization Environment). Adv. Imaging **16**(6), 56 (2001)
10. International Standards Organization (ISO): Ergonomic Requirements for Office Work with Visual Display Terminals – Part 5: Workstation Layout and Postural Requirements (1998). Ref.No.: ISO 9241-5:1998
11. International Standards Organization (ISO): Ergonomic Design of Control Centres – Part 4: Layout and Dimensions of Workstations (2004). Ref.No.: ISO 11064-4:2004
12. Rekimoto, J.: Transvision: a hand-held augmented reality system for collaborative design. In: Proceedings of Virtual Systems and Multimedia, Gifu, vol. 96, pp. 18–20 (1996)
13. Tönnis, M., Benzina, A., Klinker, G.: Utilizing consumer 3D TV hardware for a flexibly reconfigurable visualization system. Technical report, Technische Universität München (2011)

Editorial Policy

1. Volumes in the following three categories will be published in LNCSE:

i) Research monographs
ii) Tutorials
iii) Conference proceedings

Those considering a book which might be suitable for the series are strongly advised to contact the publisher or the series editors at an early stage.

2. Categories i) and ii). Tutorials are lecture notes typically arising via summer schools or similar events, which are used to teach graduate students. These categories will be emphasized by Lecture Notes in Computational Science and Engineering. **Submissions by interdisciplinary teams of authors are encouraged.** The goal is to report new developments – quickly, informally, and in a way that will make them accessible to non-specialists. In the evaluation of submissions timeliness of the work is an important criterion. Texts should be well-rounded, well-written and reasonably self-contained. In most cases the work will contain results of others as well as those of the author(s). In each case the author(s) should provide sufficient motivation, examples, and applications. In this respect, Ph.D. theses will usually be deemed unsuitable for the Lecture Notes series. Proposals for volumes in these categories should be submitted either to one of the series editors or to Springer-Verlag, Heidelberg, and will be refereed. A provisional judgement on the acceptability of a project can be based on partial information about the work: a detailed outline describing the contents of each chapter, the estimated length, a bibliography, and one or two sample chapters – or a first draft. A final decision whether to accept will rest on an evaluation of the completed work which should include

- at least 100 pages of text;
- a table of contents;
- an informative introduction perhaps with some historical remarks which should be accessible to readers unfamiliar with the topic treated;
- a subject index.

3. Category iii). Conference proceedings will be considered for publication provided that they are both of exceptional interest and devoted to a single topic. One (or more) expert participants will act as the scientific editor(s) of the volume. They select the papers which are suitable for inclusion and have them individually refereed as for a journal. Papers not closely related to the central topic are to be excluded. Organizers should contact the Editor for CSE at Springer at the planning stage, see *Addresses* below.

In exceptional cases some other multi-author-volumes may be considered in this category.

4. Only works in English will be considered. For evaluation purposes, manuscripts may be submitted in print or electronic form, in the latter case, preferably as pdf- or zipped ps-files. Authors are requested to use the LaTeX style files available from Springer at http://www.springer.com/authors/book+authors/helpdesk?SGWID=0-1723113-12-971304-0 (Click on Templates → LaTeX → monographs or contributed books).

For categories ii) and iii) we strongly recommend that all contributions in a volume be written in the same LaTeX version, preferably LaTeX2e. Electronic material can be included if appropriate. Please contact the publisher.

Careful preparation of the manuscripts will help keep production time short besides ensuring satisfactory appearance of the finished book in print and online.

5. The following terms and conditions hold. Categories i), ii) and iii):

Authors receive 50 free copies of their book. No royalty is paid.
Volume editors receive a total of 50 free copies of their volume to be shared with authors, but no royalties.

Authors and volume editors are entitled to a discount of 33.3 % on the price of Springer books purchased for their personal use, if ordering directly from Springer.

6. Springer secures the copyright for each volume.

Addresses:

Timothy J. Barth
NASA Ames Research Center
NAS Division
Moffett Field, CA 94035, USA
barth@nas.nasa.gov

Michael Griebel
Institut für Numerische Simulation
der Universität Bonn
Wegelerstr. 6
53115 Bonn, Germany
griebel@ins.uni-bonn.de

David E. Keyes
Mathematical and Computer Sciences
and Engineering
King Abdullah University of Science
and Technology
P.O. Box 55455
Jeddah 21534, Saudi Arabia
david.keyes@kaust.edu.sa

and

Department of Applied Physics
and Applied Mathematics
Columbia University
500 W. 120 th Street
New York, NY 10027, USA
kd2112@columbia.edu

Risto M. Nieminen
Department of Applied Physics
Aalto University School of Science
and Technology
00076 Aalto, Finland
risto.nieminen@aalto.fi

Dirk Roose
Department of Computer Science
Katholieke Universiteit Leuven
Celestijnenlaan 200A
3001 Leuven-Heverlee, Belgium
dirk.roose@cs.kuleuven.be

Tamar Schlick
Department of Chemistry
and Courant Institute
of Mathematical Sciences
New York University
251 Mercer Street
New York, NY 10012, USA
schlick@nyu.edu

Editor for Computational Science
and Engineering at Springer:
Martin Peters
Springer-Verlag
Mathematics Editorial IV
Tiergartenstrasse 17
69121 Heidelberg, Germany
martin.peters@springer.com

Lecture Notes in Computational Science and Engineering

1. D. Funaro, *Spectral Elements for Transport-Dominated Equations.*
2. H.P. Langtangen, *Computational Partial Differential Equations.* Numerical Methods and Diffpack Programming.
3. W. Hackbusch, G. Wittum (eds.), *Multigrid Methods V.*
4. P. Deuflhard, J. Hermans, B. Leimkuhler, A.E. Mark, S. Reich, R.D. Skeel (eds.), *Computational Molecular Dynamics: Challenges, Methods, Ideas.*
5. D. Kröner, M. Ohlberger, C. Rohde (eds.), *An Introduction to Recent Developments in Theory and Numerics for Conservation Laws.*
6. S. Turek, *Efficient Solvers for Incompressible Flow Problems.* An Algorithmic and Computational Approach.
7. R. von Schwerin, ***M**ulti **B**ody **S**ystem **SIM**ulation.* Numerical Methods, Algorithms, and Software.
8. H.-J. Bungartz, F. Durst, C. Zenger (eds.), *High Performance Scientific and Engineering Computing.*
9. T.J. Barth, H. Deconinck (eds.), *High-Order Methods for Computational Physics.*
10. H.P. Langtangen, A.M. Bruaset, E. Quak (eds.), *Advances in Software Tools for Scientific Computing.*
11. B. Cockburn, G.E. Karniadakis, C.-W. Shu (eds.), *Discontinuous Galerkin Methods.* Theory, Computation and Applications.
12. U. van Rienen, *Numerical Methods in Computational Electrodynamics.* Linear Systems in Practical Applications.
13. B. Engquist, L. Johnsson, M. Hammill, F. Short (eds.), *Simulation and Visualization on the Grid.*
14. E. Dick, K. Riemslagh, J. Vierendeels (eds.), *Multigrid Methods VI.*
15. A. Frommer, T. Lippert, B. Medeke, K. Schilling (eds.), *Numerical Challenges in Lattice Quantum Chromodynamics.*
16. J. Lang, *Adaptive Multilevel Solution of Nonlinear Parabolic PDE Systems.* Theory, Algorithm, and Applications.
17. B.I. Wohlmuth, *Discretization Methods and Iterative Solvers Based on Domain Decomposition.*
18. U. van Rienen, M. Günther, D. Hecht (eds.), *Scientific Computing in Electrical Engineering.*
19. I. Babuška, P.G. Ciarlet, T. Miyoshi (eds.), *Mathematical Modeling and Numerical Simulation in Continuum Mechanics.*
20. T.J. Barth, T. Chan, R. Haimes (eds.), *Multiscale and Multiresolution Methods.* Theory and Applications.
21. M. Breuer, F. Durst, C. Zenger (eds.), *High Performance Scientific and Engineering Computing.*
22. K. Urban, *Wavelets in Numerical Simulation.* Problem Adapted Construction and Applications.

23. L.F. Pavarino, A. Toselli (eds.), *Recent Developments in Domain Decomposition Methods.*

24. T. Schlick, H.H. Gan (eds.), *Computational Methods for Macromolecules: Challenges and Applications.*

25. T.J. Barth, H. Deconinck (eds.), *Error Estimation and Adaptive Discretization Methods in Computational Fluid Dynamics.*

26. M. Griebel, M.A. Schweitzer (eds.), *Meshfree Methods for Partial Differential Equations.*

27. S. Müller, *Adaptive Multiscale Schemes for Conservation Laws.*

28. C. Carstensen, S. Funken, W. Hackbusch, R.H.W. Hoppe, P. Monk (eds.), *Computational Electromagnetics.*

29. M.A. Schweitzer, *A Parallel Multilevel Partition of Unity Method for Elliptic Partial Differential Equations.*

30. T. Biegler, O. Ghattas, M. Heinkenschloss, B. van Bloemen Waanders (eds.), *Large-Scale PDE-Constrained Optimization.*

31. M. Ainsworth, P. Davies, D. Duncan, P. Martin, B. Rynne (eds.), *Topics in Computational Wave Propagation.* Direct and Inverse Problems.

32. H. Emmerich, B. Nestler, M. Schreckenberg (eds.), *Interface and Transport Dynamics.* Computational Modelling.

33. H.P. Langtangen, A. Tveito (eds.), *Advanced Topics in Computational Partial Differential Equations.* Numerical Methods and Diffpack Programming.

34. V. John, *Large Eddy Simulation of Turbulent Incompressible Flows.* Analytical and Numerical Results for a Class of LES Models.

35. E. Bänsch (ed.), *Challenges in Scientific Computing - CISC 2002.*

36. B.N. Khoromskij, G. Wittum, *Numerical Solution of Elliptic Differential Equations by Reduction to the Interface.*

37. A. Iske, *Multiresolution Methods in Scattered Data Modelling.*

38. S.-I. Niculescu, K. Gu (eds.), *Advances in Time-Delay Systems.*

39. S. Attinger, P. Koumoutsakos (eds.), *Multiscale Modelling and Simulation.*

40. R. Kornhuber, R. Hoppe, J. Périaux, O. Pironneau, O. Wildlund, J. Xu (eds.), *Domain Decomposition Methods in Science and Engineering.*

41. T. Plewa, T. Linde, V.G. Weirs (eds.), *Adaptive Mesh Refinement – Theory and Applications.*

42. A. Schmidt, K.G. Siebert, *Design of Adaptive Finite Element Software.* The Finite Element Toolbox ALBERTA.

43. M. Griebel, M.A. Schweitzer (eds.), *Meshfree Methods for Partial Differential Equations II.*

44. B. Engquist, P. Lötstedt, O. Runborg (eds.), *Multiscale Methods in Science and Engineering.*

45. P. Benner, V. Mehrmann, D.C. Sorensen (eds.), *Dimension Reduction of Large-Scale Systems.*

46. D. Kressner, *Numerical Methods for General and Structured Eigenvalue Problems.*

47. A. Boriçi, A. Frommer, B. Joó, A. Kennedy, B. Pendleton (eds.), *QCD and Numerical Analysis III.*

48. F. Graziani (ed.), *Computational Methods in Transport.*

49. B. Leimkuhler, C. Chipot, R. Elber, A. Laaksonen, A. Mark, T. Schlick, C. Schütte, R. Skeel (eds.), *New Algorithms for Macromolecular Simulation.*

50. M. Bücker, G. Corliss, P. Hovland, U. Naumann, B. Norris (eds.), *Automatic Differentiation: Applications, Theory, and Implementations.*

51. A.M. Bruaset, A. Tveito (eds.), *Numerical Solution of Partial Differential Equations on Parallel Computers.*

52. K.H. Hoffmann, A. Meyer (eds.), *Parallel Algorithms and Cluster Computing.*

53. H.-J. Bungartz, M. Schäfer (eds.), *Fluid-Structure Interaction.*

54. J. Behrens, *Adaptive Atmospheric Modeling.*

55. O. Widlund, D. Keyes (eds.), *Domain Decomposition Methods in Science and Engineering XVI.*

56. S. Kassinos, C. Langer, G. Iaccarino, P. Moin (eds.), *Complex Effects in Large Eddy Simulations.*

57. M. Griebel, M.A Schweitzer (eds.), *Meshfree Methods for Partial Differential Equations III.*

58. A.N. Gorban, B. Kégl, D.C. Wunsch, A. Zinovyev (eds.), *Principal Manifolds for Data Visualization and Dimension Reduction.*

59. H. Ammari (ed.), *Modeling and Computations in Electromagnetics: A Volume Dedicated to Jean-Claude Nédélec.*

60. U. Langer, M. Discacciati, D. Keyes, O. Widlund, W. Zulehner (eds.), *Domain Decomposition Methods in Science and Engineering XVII.*

61. T. Mathew, *Domain Decomposition Methods for the Numerical Solution of Partial Differential Equations.*

62. F. Graziani (ed.), *Computational Methods in Transport: Verification and Validation.*

63. M. Bebendorf, *Hierarchical Matrices.* A Means to Efficiently Solve Elliptic Boundary Value Problems.

64. C.H. Bischof, H.M. Bücker, P. Hovland, U. Naumann, J. Utke (eds.), *Advances in Automatic Differentiation.*

65. M. Griebel, M.A. Schweitzer (eds.), *Meshfree Methods for Partial Differential Equations IV.*

66. B. Engquist, P. Lötstedt, O. Runborg (eds.), *Multiscale Modeling and Simulation in Science.*

67. I.H. Tuncer, Ü. Gülcat, D.R. Emerson, K. Matsuno (eds.), *Parallel Computational Fluid Dynamics 2007.*

68. S. Yip, T. Diaz de la Rubia (eds.), *Scientific Modeling and Simulations.*

69. A. Hegarty, N. Kopteva, E. O'Riordan, M. Stynes (eds.), *BAIL* 2008 – *Boundary and Interior Layers.*

70. M. Bercovier, M.J. Gander, R. Kornhuber, O. Widlund (eds.), *Domain Decomposition Methods in Science and Engineering XVIII.*

71. B. Koren, C. Vuik (eds.), *Advanced Computational Methods in Science and Engineering.*

72. M. Peters (ed.), *Computational Fluid Dynamics for Sport Simulation.*

73. H.-J. Bungartz, M. Mehl, M. Schäfer (eds.), *Fluid Structure Interaction II - Modelling, Simulation, Optimization.*

74. D. Tromeur-Dervout, G. Brenner, D.R. Emerson, J. Erhel (eds.), *Parallel Computational Fluid Dynamics 2008.*

75. A.N. Gorban, D. Roose (eds.), *Coping with Complexity: Model Reduction and Data Analysis.*

76. J.S. Hesthaven, E.M. Rønquist (eds.), *Spectral and High Order Methods for Partial Differential Equations.*

77. M. Holtz, *Sparse Grid Quadrature in High Dimensions with Applications in Finance and Insurance.*

78. Y. Huang, R. Kornhuber, O.Widlund, J. Xu (eds.), *Domain Decomposition Methods in Science and Engineering XIX.*

79. M. Griebel, M.A. Schweitzer (eds.), *Meshfree Methods for Partial Differential Equations V.*

80. P.H. Lauritzen, C. Jablonowski, M.A. Taylor, R.D. Nair (eds.), *Numerical Techniques for Global Atmospheric Models.*

81. C. Clavero, J.L. Gracia, F.J. Lisbona (eds.), *BAIL 2010 – Boundary and Interior Layers, Computational and Asymptotic Methods.*

82. B. Engquist, O. Runborg, Y.R. Tsai (eds.), *Numerical Analysis and Multiscale Computations.*

83. I.G. Graham, T.Y. Hou, O. Lakkis, R. Scheichl (eds.), *Numerical Analysis of Multiscale Problems.*

84. A. Logg, K.-A. Mardal, G. Wells (eds.), *Automated Solution of Differential Equations by the Finite Element Method.*

85. J. Blowey, M. Jensen (eds.), *Frontiers in Numerical Analysis - Durham 2010.*

86. O. Kolditz, U.-J. Gorke, H. Shao, W. Wang (eds.), *Thermo-Hydro-Mechanical-Chemical Processes in Fractured Porous Media - Benchmarks and Examples.*

87. S. Forth, P. Hovland, E. Phipps, J. Utke, A. Walther (eds.), *Recent Advances in Algorithmic Differentiation.*

88. J. Garcke, M. Griebel (eds.), *Sparse Grids and Applications.*

89. M. Griebel, M.A. Schweitzer (eds.), *Meshfree Methods for Partial Differential Equations VI.*

90. C. Pechstein, *Finite and Boundary Element Tearing and Interconnecting Solvers for Multiscale Problems.*

91. R. Bank, M. Holst, O. Widlund, J. Xu (eds.), *Domain Decomposition Methods in Science and Engineering XX.*

92. H. Bijl, D. Lucor, S. Mishra, C. Schwab (eds.), *Uncertainty Quantification in Computational Fluid Dynamics.*

93. M. Bader, H.-J. Bungartz, T. Weinzierl (eds.), *Advanced Computing.*

For further information on these books please have a look at our mathematics catalogue at the following URL: www.springer.com/series/3527

Monographs in Computational Science and Engineering

1. J. Sundnes, G.T. Lines, X. Cai, B.F. Nielsen, K.-A. Mardal, A. Tveito, *Computing the Electrical Activity in the Heart.*

For further information on this book, please have a look at our mathematics catalogue at the following URL: www.springer.com/series/7417

Texts in Computational Science and Engineering

1. H. P. Langtangen, *Computational Partial Differential Equations.* Numerical Methods and Diffpack Programming. 2nd Edition
2. A. Quarteroni, F. Saleri, P. Gervasio, *Scientific Computing with MATLAB and Octave.* 3rd Edition
3. H. P. Langtangen, *Python Scripting for Computational Science.* 3rd Edition
4. H. Gardner, G. Manduchi, *Design Patterns for e-Science.*
5. M. Griebel, S. Knapek, G. Zumbusch, *Numerical Simulation in Molecular Dynamics.*
6. H. P. Langtangen, *A Primer on Scientific Programming with Python.* 3rd Edition
7. A. Tveito, H. P. Langtangen, B. F. Nielsen, X. Cai, *Elements of Scientific Computing.*
8. B. Gustafsson, *Fundamentals of Scientific Computing.*
9. M. Bader, *Space-Filling Curves.*
10. M. Larson, F. Bengzon, *The Finite Element Method: Theory, Implementation and Applications.*

For further information on these books please have a look at our mathematics catalogue at the following URL: www.springer.com/series/5151

MIX
Papier aus verantwortungsvollen Quellen
Paper from responsible sources
FSC® C105338

If you have any concerns about our products,
you can contact us on
ProductSafety@springernature.com

In case Publisher is established outside the EU,
the EU authorized representative is:
Springer Nature Customer Service Center GmbH
Europaplatz 3, 69115 Heidelberg, Germany

Printed by Libri Plureos GmbH
in Hamburg, Germany